Springer Theses

Recognizing Outstanding Ph.D. Research

More information about this series at http://www.springer.com/series/8790

Aims and Scope

The series "Springer Theses" brings together a selection of the very best Ph.D. theses from around the world and across the physical sciences. Nominated and endorsed by two recognized specialists, each published volume has been selected for its scientific excellence and the high impact of its contents for the pertinent field of research. For greater accessibility to non-specialists, the published versions include an extended introduction, as well as a foreword by the student's supervisor explaining the special relevance of the work for the field. As a whole, the series will provide a valuable resource both for newcomers to the research fields described, and for other scientists seeking detailed background information on special questions. Finally, it provides an accredited documentation of the valuable contributions made by today's younger generation of scientists.

Theses are accepted into the series by invited nomination only and must fulfill all of the following criteria

- They must be written in good English.
- The topic should fall within the confines of Chemistry, Physics, Earth Sciences, Engineering and related interdisciplinary fields such as Materials, Nanoscience, Chemical Engineering, Complex Systems and Biophysics.
- The work reported in the thesis must represent a significant scientific advance.
- If the thesis includes previously published material, permission to reproduce this must be gained from the respective copyright holder.
- They must have been examined and passed during the 12 months prior to nomination.
- Each thesis should include a foreword by the supervisor outlining the significance of its content.
- The theses should have a clearly defined structure including an introduction accessible to scientists not expert in that particular field.

Marc Z. Miskin

The Automated Design of Materials Far From Equilibrium

Doctoral Thesis accepted by the University of Chicago, Chicago, Illinois, USA

Springer

Marc Z. Miskin
Cornell University
Ithaca, NY, USA

ISSN 2190-5053 ISSN 2190-5061 (electronic)
Springer Theses
ISBN 978-3-319-24619-2 ISBN 978-3-319-24621-5 (eBook)
DOI 10.1007/978-3-319-24621-5

Library of Congress Control Number: 2015952781

Springer Cham Heidelberg New York Dordrecht London

Printed on acid-free paper

Springer International Publishing AG Switzerland is part of Springer Science+Business Media (www.springer.com)

"Our guiding principle was that design is neither an intellectual nor a material affair, but simply an integral part of the stuff of life…" – Walter Gropius

For Pop

Supervisor's Foreword

We often think about materials by considering two domains: micro-level of constituent components and macro-level of overall material properties. The kinds of things considered as constituent components can vary in scale, ranging from atoms in a crystal to molecules in a polymer chain to millimeter-size particles in a granular material, but through their interactions, they collectively determine the macro-level behavior of the material. This behavior in turn determines whether the material is interesting for science and useful for technology.

By materials design, we are envisioning a process that gets us from desired overall properties to requirements for the constituent components. In such a paradigm, information must flow from the top down, starting at the desired macro-level result to end at the micro-level cause that produces it. This inverse process stands in contrast to the typical relationships built up by physics and materials science, which start at the microscopic roots and work up to describe bulk behavior.

Achieving targeted materials properties through rational design is undoubtedly a highly exciting and promising goal. However, it is also complicated due to at least four intrinsic problems.

The first stems from the fact that tuning a material's behavior often requires a large number of control parameters. Yet as the number of control parameters increases, the set of possible combinations grows exponentially. Thus, the number of possible micro-level configurations to parse in searching for desirable behavior can quickly become inexhaustible.

Second, the mapping from macro-level behavior to micro-level features may not be known, and even if it is, it may not be unique. In other words, even when it is possible to compute a macroscopic response for a given microscale configuration, there may be many other microscale configurations that lead to the same properties. This makes it unclear how one can systematically or rationally parse through potentially promising material configurations.

Third, for many of the most interesting and certainly for the more complex materials, we have only a very rudimentary understanding of the underlying physics. Yet these often are precisely the materials we would want to optimize. The question then becomes how one can approach efficient design in the absence of a detailed descriptive model, perhaps relying only on experiments and computer simulations.

Finally, to make a design-based solution valuable, it should do more than represent a single answer to a single inverse problem. What would truly enable materials design is a system that provides a path toward solving similar problems without having to go through the full inverse process again. Ultimately, we would like to obtain more general outcomes, or design rules, that apply to whole classes of related inverse problems.

This thesis develops a new methodology to tackle the inverse process that is at the heart of materials design, and it applies this new approach to materials far from equilibrium, a frontier of materials research scientifically and across many industrial applications. A prototype of such materials are jammed granular aggregates. In their most generic form, these are amorphous collections of identical particles that collectively act as a rigid solid when pressing against one another. They represent an excellent test case for the task of materials design, because despite their seemingly simple nature, they exhibit highly complex behaviors and emphasize all the difficulties of design listed above; except for a few limited cases, there is no general predictive model that can map the overall bulk behavior to particle level aspects such as shape; shape furthermore is an effectively inexhaustible parameter, and the disordered, far-from-equilibrium nature of jammed granular aggregates implies that the standard methods for finding optimized particle configurations, as developed for ordered crystalline arrangements, no longer apply.

A central innovation in this thesis is to approach design problems involving jammed aggregates by taking a cue from biology. Given candidate design solutions for this type of problem, the question which candidate performs better than another at the task at hand can typically be answered only by evaluating the performance with either computer simulation or direct experiment. What this thesis shows is that design can proceed by discarding the worse candidate and replacing it with a small variation, or mutation, of the better candidate. This process forms the kernel of a class of optimization algorithms known as evolutionary algorithms. In turn, these algorithms form the backbone for an automated approach to the design of jammed matter.

Each of the four key problems surrounding design can be readily addressed using this formalism. Evolutionary optimization provides a systematic way to parse large parameter spaces. Indeed in the work presented here, it can be used to design the particle shapes in a jammed aggregate when the dimension of the search space is as large as 40. Moreover, the process works for problems that have multiple solutions and when the relationship between input and output is unknown, the only real requirement being a measure of quality to compare potential problem solutions. Finally, this thesis takes some of the first strides toward extending the results of a single optimization to be used as "design rules." In particular, it presents rules that can be used by themselves to find a particle shape that packs to any target density.

The perspective that jammed aggregates can be a "designer material" has led to new insights in both how to control a jammed material and also why it behaves the way it does. For instance, this thesis identifies particle shapes that form the stiffest, the softest, the densest, and the loosest aggregates. In each case, the shapes reveal not only new opportunities to use jammed aggregates as enabling technologies, but the results themselves develop a picture about what is key in setting jammed behavior: interlocking particles promote stiff packings and rod-like particles subject to torque produce loose, soft packings. It also presents exotic results, such as a particle that when packed in aggregate stiffens under compression. This behavior is rather atypical for granular matter and shows how design paradigms can facilitate the discovery of new behavior, outside the envelope of known responses.

While the bulk of the work in this thesis is focused on jammed aggregates, the issues at the heart of it are much broader. True design that starts from desired macro-level materials properties to identify the required micro-level ingredients has only very recently come into reach, mainly because of increased computational power but also because of new algorithmic approaches. The particular approach developed here for designing jammed aggregates is not based on a specific material or on a specific size scale and as such, is likely to find applications across a wide array of different systems, ranging from granular materials to colloids to polymers to nanoparticles. Thus, it opens up an exciting new direction for materials design as a whole.

Chicago, USA
September 1, 2015

Heinrich M. Jaeger

Preface

Automated design is emerging as a powerful concept in materials science. By combining computer algorithms, simulations, and experimental data, new techniques are being developed that start with high-level functional requirements and identify the ideal materials that achieve them. This represents a radically different picture of how materials become functional in which technological demand drives material discovery, rather than the other way around.

At the frontiers of this field, materials systems previously considered too complicated can start to be controlled and understood. Particularly promising are materials far from equilibrium. Material robustness, high strength, self-healing, and memory are properties displayed by several materials systems that are intrinsically out of equilibrium. These and other properties could be revolutionary, provided they can first be controlled.

This thesis conceptualizes and implements a framework for designing materials that are far from equilibrium. We show how, even in the absence of a complete physical theory, design from the top down is possible and lends itself to producing physical insight. As a prototype system, we work with granular materials: collections of athermal, macroscopic identical objects, since these materials function both as an essential component of industrial processes as well as a model system for many non-equilibrium states of matter. We show that by placing granular materials in the context of design, benefits emerge simultaneously for fundamental and applied interests.

As first steps, we use our framework to design granular aggregates with extreme properties like high stiffness and softness. We demonstrate control over nonlinear effects by producing exotic aggregates that stiffen under compression. Expanding on our framework, we conceptualize new ways of thinking about material design when automatic discovery is possible. We show how to build rules that link particle shapes to arbitrary granular packing density. We examine how the results of a design process are contingent upon operating conditions by studying which shapes dissipate energy fastest in a granular gas. We even move to create optimization algorithms for the expressed purpose of material design, by integrating them with statistical mechanics. In all of these cases, we show that turning to machines puts a fresh perspective on materials far from equilibrium. By matching forms to functions, complexities become possibilities, motifs emerge that describe new physics, and the door opens to rational design.

Ithaca, NY, USA Marc Z. Miskin

Acknowledgments

I have always been proud of working in the James Franck Institute as a whole. It's a very special place because of the fact that it is a community of researchers who share ideas and encourage one another. That said virtually everybody at the JFI deserves some kind of acknowledgement for creating an atmosphere that encourages new ideas and collaboration.

A few people in particular, though, were integral for me to build my own ideas into something new. The most important, without a doubt, is my advisor Heinrich Jaeger. Heinrich is responsible for several deep changes in my thinking about how science relates with society and what it means to do science. The basic notion of how applications interplay with fundamental research simultaneously underscores this thesis, and I think he is responsible for that change in perspective. More than the philosophical, he has been there to encourage me when I was stuck in the hard parts of the work, he constantly worked to provide the tools I've needed, and he gave me probably more freedom than I deserved in going forward. As a whole, it's been very special to build up this work with him, sharing in the ideas and the directions, and I can't thank him enough.

A few key members of the Jaeger group and JFI also need to be acknowledged. In particular, Scott Waitukaitis, who was my lab mate for 4 years and a good friend, was key in getting me plugged into the lab in the first place, giving sagely advice and explaining the merits of manhandling data. Eric Brown and Justin Burton also deserve thanks for teaching me a lot about the real-life gritty aspects of getting research done. They were also independently two of the most fun postdocs you could hope for as a first year grad student. Tom Witten deserves some serious thanks. If you want to get a new idea off the ground, Tom is exactly the person you need to have around. He is always able to see what's cool in what you're doing, even if you're not, and I needed that kind of support to keep going. Finally, I have to give major thanks to my thesis committee of Jeff Harvey, Sidney Nagel, and Wendy Zhang. All three have been abundantly courteous, accommodating, and constructive throughout the process of creating my thesis.

Outside the JFI, Chicago has also been a great place for me in terms of friends. I was lucky enough to be in a graduate program with several guys who turned out to be great friends: Bobby Lanza, Jake Ellowitz, Sam Meehan, and Jordan Webster. In particular, Jake has also been a great colleague who helped vet and invent several of the ideas in this work (Fig. 3.1c was totally his idea).

Last, there is my family who is always essential. My parents always provided the support I needed to even reach this point. My grandfather, to whom the thesis is dedicated, deserves the thanks for inspiring my interest in science in the first place. And, finally, there is my fiance, Monika, who supports me constantly in everything that I do. She's responsible for always providing me with encouragement, love, and support (and at times actually helping in the lab). She certainly deserves thanks for not just keeping my work on the right track, but most of the rest of my life too. I'm profoundly happy and lucky to have her with me through this process and in everything that comes next.

Contents

Chapter 1
Introduction

The automatic design of materials is emerging as field in the twenty-first century (Jain et al. 2013; Hautier et al. 2012; Curtarolo et al. 2013; Oganov et al. 2011; Dahiyat and Mayo 1997). Its ultimate goal is a system that take functional requirements as input and produces the materials that achieve them as output. In other words, this new field seeks a systematic framework that starts with large scale macroscopic properties and finishes by connecting them to a microscopic configuration with those properties.

This process, going from macroscopic phenomena to microscopic features, is what we call design. It represents a markedly different picture of how to move from physical concepts to useful materials. Materials would no longer be discovered by accident nor appropriated from nature. Instead, the ideal material, as defined by precise and restrictive functionalities, could be cooked up to meet increasingly complicated technological demands. Already this concept has shown promise in designing Lithium ion battery cathodes (Chen et al. 2012), photocatalyists (Castelli et al. 2012), and steel alloys (Olson 1997) from the top-down.

Success in this endeavor requires learning to use physical theories as not just predictive tools, but as design tools. The power to calculate macroscopic properties from microscopic relationships is a major strength of physics. But even armed with this knowledge, it is unclear how to generally address the reverse. For example, even if it were possible to clearly predict the strength of a metal based on a given arrangement of its atoms, it would still be extremely difficult to answer what arrangement of atoms maximizes strength. Generally speaking, the capacity to predict properties given microscopic features does not intrinsically yield the power to select the microscopic features that produce desired properties.

Arguably, the difficulties and benefits associated with transforming predictions to designs are typified by materials far from equilibrium. These materials can exhibit extreme properties, desirable for new applications. For instance, rapidly cooled metal alloys form some of the strongest materials known to man (Inoue 2000; Wang et al. 2004). Far from equilibrium is where self-healing

M.Z. Miskin, *The Automated Design of Materials Far From Equilibrium*,
Springer Theses, DOI 10.1007/978-3-319-24621-5_1

(Wool 2008; Nosonovsky et al. 2009) and memory materials (Gunes and Jana 2008) are often found. Finally, some of the most intricate patterns formed in nature occur when conditions are driven away from the steady state (Cross and Hohenberg 1993; Langer 1980; Lopes and Jaeger 2001). In short, there are extraordinary properties associated with materials beyond equilibrium, and several of which hold promise for transformative technologies.

Conversely, designing these materials has proven notoriously difficult, ironically because of the complicated interactions that make them so intriguing. For example, boundary conditions and evolution equations are often not enough to make predictions. Instead, full processing conditions often have to specified as far from equilibrium systems can possess memory properties (Josserand et al. 2000; Bianchi et al. 2010). Furthermore, dissipation cannot be treated perturbatively. Instead, it often plays a crucial role, especially in structure formation for continually driven systems (Cross and Hohenberg 1993; Metcalf et al. 1997). And while atoms and planets can easily be approximated as points or spheres, constituent shapes can play important, if not prominent, roles for many processes outside the equilibrium regime (Glotzer and Solomon 2007; Athanassiadis et al. 2014).

To the extent that far from equilibrium materials represent a unique class of materials poised to solve contemporary, technological problems, a method must exist that addresses how to design them using the limited scope of physical knowledge and accounting for all their intrinsic complexities. This thesis builds a framework that meets that demand. Furthermore, the result can be employed to solve less complicated and more general design problems, without substantial alteration. In short, we take some of the first steps towards an automatic framework for materials discovery that works even for systems far from equilibrium.

As a prototype system, we mostly focus on granular materials. At face value, granular materials are aggregates of large, athermal objects that interact via short range repulsion (Jaeger et al. 1996; Duran 1999; Kadanoff 1999). Yet many of the features that make far from equilibrium materials intriguing and complicated present themselves. Granular materials display self-healing, memory of loading protocols (Josserand et al. 2000; Majmudar and Behringer 2005), and distribute forces over a range of scales (Mueth et al. 1998). They can resist load like solids, but are rarely linearly elastic (Athanassiadis et al. 2014). They can be driven to behave like gases, but due to dissipation, they often violate basic thermodynamic concepts like equipartition (Feitosa and Menon 2002) and homogeneity (Eggers 1999). Ultimately, the design of granular materials is promising, with recent applications to robotics (Brown et al. 2010) and architecture (Dierichs and Menges 2012), but also complicated by the fact that they are virtually always disordered, particle shape impacts almost all of their measurable properties, and that there is no predictive theory that generally explains their behavior.

In the absence of a general theory, any design process must operate using only the power to comparatively evaluate potential solutions. For example, one could use simulations to compute the properties of a given granular material and compare whether one design is better than another without having to address why. This is sufficient for a design process of generate and evaluate, provided the former adapts to the results of the later.

Written this way, the task at hand is reminiscent of biological evolution. In fact, complicated search and optimization problems, ranging from robot morphology design (Lipson and Pollack 2000) to crystal structure prediction (Oganov et al. 2011), have already been solved by building computer algorithms around the basic idea of evolution (Eiben and Smith 2003). Given a population of possible designs, simulations allow the best performing objects to be determined. New, potentially better objects could be generated by randomly mutating them. To the extent that iterating this process leads to improvement, objects can be brought progressively closer to solving a design goal. Thus, artificial evolution can, in principle, be used as the missing link that transforms the predictive power in simulations into a tool that links macro to micro.

With artificial evolution as a starting point, the next steps can be taken to building a system for automated materials design far from equilibrium. We feel it is compelling to address problems where comparatively little prior work exits, presumably owing to the problem complexity. Furthermore, we have attempted to produce proofs of principle for solving general classes of design problems with an evolution framework. The design problems considered are:

Optimization: Find microscopic configurations that extremize a user specified macroscopic property. We specifically consider the task of finding which granular particle shapes maximize/minimize stiffness under compression and discover a granular particle that produces extreme nonlinear behavior by stiffening while strained.

Inverse Problems: For a given material property, find a means to rationally link arbitrary responses to each specific microscopic configuration that generates them. We build an inverse problem solution that links at least one granular shape to any user specified volume fraction in an optimized range.

Transitions of Designs: If a design goal is kept the same, but the conditions the material is subject to are varied, when and why do the qualitative features of an optimization change? We discover the shapes that are first to freeze in a granular gas at a fixed initial density. By varying that density, we are able to examine the stability of solutions as well as extract qualitative physics about how the granular gas cools.

Online Design: Could a computer algorithm actively interact with an experiment in silico to optimize a material? We create a simple algorithm that uses statistical mechanics to incrementally adapt materials to design requirements. We task the algorithm with morphing energy landscapes to control the location of a thermalized particle and folding a chain of six particles into an octahedron.

In each of these four cases, we use evolution to quickly generate an effective solution. For first two granular material problems, we also show that the resulting designs can be easily fabricated and evaluated in the real world with 3D printing. The complete paradigm allows one to go from a design concept to a functioning prototype in the timescale of weeks without ever having to produce a reductive physical theory.

The remainder of this thesis is organized as follows. First, we discuss the theoretical foundations of artificial evolution, both from a computer science perspective and from a biological one. The former addresses how artificial evolution works, whereas the later address why. With this as a general purpose framework for optimization, we show how to link each of the enumerated design problems with an optimization problem and artificial evolution is then implemented to produce solutions. Repeatedly, the results are synthesized to yield new scientific insight about physics far from equilibrium. We conclude each section by showing how each case forms a proof of principle for broadly solving each class of design problems, even beyond equilibrium. Specifically, we show how generalizations of the problems solved here are straightforward and reduce the complexity of the design task. To the extent that we are attacking some of the hardest problems in materials design, the evolution framework stands poised to design numerous simpler materials. In the final chapter, we put the results into perspective by addressing what it means to design, how that task relates to the goals of science, and what risks and rewards are associated with artificial evolution.

References

Athanassiadis, A. G., Miskin, M. Z., Kaplan, P., Rodenberg, N., Lee, S. H., Merritt, J., Brown, E., Amend, J., Lipson, H., & Jaeger, H. M. (2014). Particle shape effects on the stress response of granular packings. *Soft Matter, 10*(1), 48–59.

Bianchi, M., Scarpa, F., & Smith, C. W. (2010). Shape memory behavior in auxetic foams: Mechanical properties. *Acta Materialia, 58*(3), 858–865.

Brown, E., Rodenberg, N., Amend, J., Mozeika, A., Steltz, E., Zakin, M. R., Lipson, H., & Jaeger, H. M. (2010). Universal robotic gripper based on the jamming of granular material. *Proceedings of the National Academy of Sciences, 107*(44), 18809–18814.

Castelli, I. E., Olsen, T., Datta, S., Landis, D. D., Dahl, S., Thygesen, K. S., & Jacobsen, K. W. (2012). Computational screening of perovskite metal oxides for optimal solar light capture. *Energy and Environmental Science, 5*(2), 5814–5819.

Chen, H., Hautier, G., Jain, A., Moore, C., Kang, B., Doe, R., Wu, L., Zhu, Y., Tang, Y., & Ceder, G. (2012). Carbonophosphates: A new family of cathode materials for li-ion batteries identified computationally. *Chemistry of Materials, 24*(11), 2009–2016.

Cross, M. C., & Hohenberg, P. C. (1993). Pattern formation outside of equilibrium. *Reviews of Modern Physics, 65*(3), 851.

Curtarolo, S., Hart, G. L., Nardelli, M. B., Mingo, N., Sanvito, S., & Levy, O. (2013). The high-throughput highway to computational materials design. *Nature Materials, 12*(3), 191–201.

Dahiyat, B. I., & Mayo, S. L. (1997). De novo protein design: Fully automated sequence selection. *Science, 278*(5335), 82–87.

Dierichs, K., & Menges, A. (2012). Aggregate structures: Material and machine computation of designed granular substances. *Architectural Design, 82*(2), 74–81.

Duran, J. (1999). *Sands, powders, and grains: An introduction to the physics of granular materials*. Berlin: Springer.

Eggers, J. (1999). Sand as maxwell's demon. *Physical Review Letters, 83*(25), 5322.

Eiben, A. E., & Smith, J. E. (2003). *Introduction to evolutionary computing*. New York: Springer.

Feitosa, K., & Menon, N. (2002). Breakdown of energy equipartition in a 2d binary vibrated granular gas. *Physical Review Letters, 88*(19), 198301.

Glotzer, S. C., & Solomon, M. J. (2007). Anisotropy of building blocks and their assembly into complex structures. *Nature Materials, 6*(8), 557–562.

Gunes, I. S., & Jana, S. C. (2008). Shape memory polymers and their nanocomposites: A review of science and technology of new multifunctional materials. *Journal of Nanoscience and Nanotechnology, 8*(4), 1616–1637.

Hautier, G., Jain, A., & Ong, S. P. (2012). From the computer to the laboratory: Materials discovery and design using first-principles calculations. *Journal of Materials Science, 47*(21), 7317–7340.

Inoue, A. (2000). Stabilization of metallic supercooled liquid and bulk amorphous alloys. *Acta Materialia, 48*(1), 279–306.

Jaeger, H. M., Nagel, S. R., & Behringer, R. P. (1996). Granular solids, liquids, and gases. *Reviews of Modern Physics, 68*(4), 1259–1273.

Jain, A., Ong, S. P., Hautier, G., Chen, W., Richards, W. D., Dacek, S., Cholia, S., Gunter, D., Skinner, D., Ceder, G., et al. (2013). The materials project: A materials genome approach to accelerating materials innovation. *APL Materials, 1*(1), 011002.

Josserand, C., Tkachenko, A. V., Mueth, D. M., & Jaeger, H. M. (2000). Memory effects in granular materials. *Physical Review Letters, 85*(17), 3632.

Kadanoff, L. P. (1999). Built upon sand: Theoretical ideas inspired by granular flows. *Reviews of Modern Physics, 71*(1), 435.

Langer, J. (1980). Instabilities and pattern formation in crystal growth. *Reviews of Modern Physics, 52*(1), 1.

Lipson, H., Pollack, J. B. (2000). Automatic design and manufacture of robotic lifeforms. *Nature, 406*(6799), 974–978.

Lopes, W. A., & Jaeger, H. M. (2001). Hierarchical self-assembly of metal nanostructures on diblock copolymer scaffolds. *Nature, 414*(6865), 735–738.

Majmudar, T. S., & Behringer, R. P. (2005). Contact force measurements and stress-induced anisotropy in granular materials. *Nature, 435*(7045), 1079–1082.

Metcalf, T. H., Knight, J. B., & Jaeger, H. M. (1997). Standing wave patterns in shallow beds of vibrated granular material. *Physica A, 236*(3–4), 202–210.

Mueth, D. M., Jaeger, H. M., & Nagel, S. R. (1998). Force distribution in a granular medium. *Physical Review E, 57*(3), 3164.

Nosonovsky, M., Amano, R., Lucci, J. M., & Rohatgi, P. K. (2009). Physical chemistry of self-organization and self-healing in metals. *Physical Chemistry Chemical Physics, 11*(41), 9530–9536.

Oganov, A. R., Lyakhov, A. O., & Valle, M. (2011). How evolutionary crystal structure prediction works-and why. *Accounts of Chemical Research, 44*(3), 227–237.

Olson, G. B. (1997). Computational design of hierarchically structured materials. *Science, 277*(5330), 1237–1242.

Wang, W.-H., Dong, C., & Shek, C. (2004). Bulk metallic glasses. *Materials Science and Engineering: R: Reports, 44*(2), 45–89.

Wool, R. P. (2008). Self-healing materials: A review. *Soft Matter, 4*(3), 400–418.

Chapter 2
Artificial Evolution

Abstract Solving design problems in materials science requires a general framework that connects target behaviors to the microscopic parameters that produce them. Here we argue why artificial evolution is a particularly attractive means to this end. First, we present a brief survey of evolutionary computation with particular emphasis on how concepts have been refined to produce state of the art optimization schemes. We present two derivations of contemporary evolutionary optimizers: one from a heuristic, computational point of view and the other biologically motivated. Finally, we use the latter to examine new, deep connections between evolutionary optimization, game theory, and information geometry.

2.1 Introduction

Evolution is arguably the simplest picture of a design. It provides a flexible framework to create objects with desired behavior, without requiring explicit problem knowledge. One simply starts with a collection (or population) of possible solutions, and determines how well each candidate solves the problem. A potentially better population can then be generated by taking the some of the best performing objects and manipulating them to produce new candidates (Fig. 2.1).

Transforming this concept into an algorithm requires some careful choices about what is meant by mutation, best members, or even a population. In fact, picking one definition against another is what sets the boundaries between different classes of evolutionary algorithms (Eiben and Smith 2003). Yet once these terms are defined, the basic structure of diversification and selection operators acting on a population is always enforced. Furthermore, the key to a successful implementation is often maintaining a balance of power between these two conflicting forces.

To show how these concepts have been progressively refined, we present a brief, historical survey of evolutionary computation. We begin with the algorithms most directly inspired by biology: genetic algorithms (Holland 1975, 1973; de Jong 1975). As originally envisioned, candidate solutions are represented as strings of bits, meant as models for DNA. Mutation takes place by flipping bits, or permuting substrings. When fit objects are selected to propagate, they reproduce

M.Z. Miskin, *The Automated Design of Materials Far From Equilibrium*,
Springer Theses, DOI 10.1007/978-3-319-24621-5_2

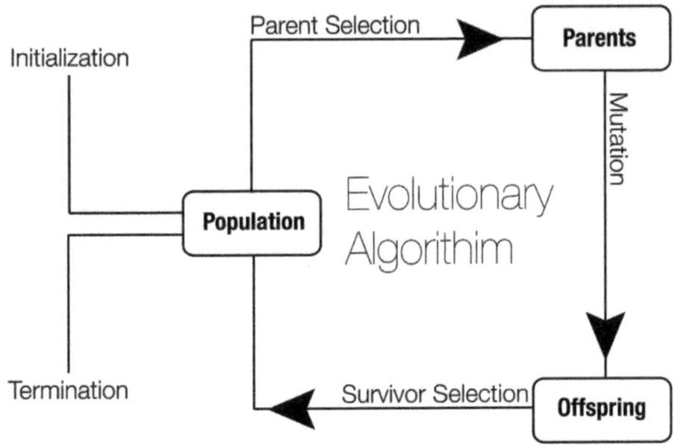

Fig. 2.1 Evolutionary algorithms abstracted to a flow-chart

recombinantly. That is, bit string fragments are taken from two or more parents and joined together to form an offspring. Altogether these algorithms are called canonical or simple genetic algorithms (Eiben and Smith 2003).

When performing optimizations in continuous spaces, some of these choices can become problematic (Hoffmeister and Back 1991). In particular, issues arise when parameter locality is not explicitly enforced (Rothlauf 2006). For example, the innocuous transformation of the binary string 0001 to 1000 corresponds to moving almost an order of magnitude in distance to the origin when converted directly to a base 10 number. Alternatively, joining two bit strings together in an ad hoc fashion can lead to offspring that lack resemblance to their parents. These large mutations can potentially overpower the effects of selection and undermine the algorithm's effectiveness. In short, the mapping from bit strings to continuous parameters can obscure the relationship between the fitness of parents and the fitness of offspring.

An obvious solution to this problem is to abandon both the bit string concept and the use of recombination. In this case, mutation becomes the sole diversifying operator and it acts directly on the variables fed to the fitness function. This picture evolution takes a collection of μ particles and subjects them each to a random modification, such as a kick of Gaussian noise. The most fit λ objects following the mutation are then copied to rebuild the μ sized population, while the least fit are discarded. Developed roughly simultaneously with genetic algorithms, these approaches are termed (λ, μ) evolution strategies (Schwefel 1993; Beyer and Schwefel 2002; Back et al. 1991).

The efficiency of an evolution strategy essentially reduces to how frequently the mutations generate good, new guesses. An ideal approach would learn to correlate the likelihood of mutations with increases in fitness. Early evolution strategies achieved this goal by introducing mutation variables as part of a candidate object (Beyer and Schwefel 2002). Effectively this allows evolution to take place in "phase space" rather than "configuration space". For example, a candidate object could be represented by a mean vector of input parameters plus the variance for a Gaussian noise term. An evolution cycle would then consist of randomly perturbing the variance, using the new variance to generate noise, and then adding that noise to the mean vector. This final object is evaluated, allowing information about the effectiveness of the mutation settings to be assessed indirectly.

This approach has several attractive features, including proofs of global convergence (Back et al. 1993) for some variants. Moreover, evolution strategies tend to converge faster and more reliably than simple genetic algorithms on a number of continuous parameter test problems (Hoffmeister and Back 1991). However, the indirect mutation parameter updates and the particle representation mean that populations often have to be large, typically consisting of hundreds of individuals. This can make evolution strategies burdensome or impractical for optimization problems where fitness evaluations correspond to expensive computations.

The computational demand associated with artificial evolution can be reduced by orders of magnitude using a more structured approach (Hansen et al. 2003; Wierstra et al. 2008). Rather than defining a population as collection of particles, one could use a probability distribution with a fixed form. With this simplification, adaptation is defined by changing in the distribution parameters. For example, if the population is fixed to be Gaussian, the parameters to adapt are the mean and the covariance matrix. The process of selection becomes tantamount to deciding how information from points sampled should be integrated to adapt these parameters.

Algorithms built around this philosophy represent the state-of-the-art for evolutionary computation. On both test bed and real world problems, methods from this family boat impressive performance and in many cases outperform algorithms like simulated annealing, particle swarm optimization, and the other evolutionary approaches (Hansen et al. 2009). The most impressive algorithm built around this philosophy is the covariance matrix adaptation evolution strategy or the CMA-ES (Hansen et al. 2003).

Figures 2.2 and 2.3 show direct comparisons between adaptive simulated annealing (Ingber et al. 2012) (ASA) and the CMA-ES on three different two dimensional test problems. Each test problem presents a different difficulty that a good black-box optimizer should be able to overcome: the first is non-separable, the second is non-convex and ill-conditioned, and the third is multimodal. An advantage was given to ASA in that its parameters were tuned first to produce quality performance whereas the CMA-ES was run with default parameters only. Yet in all three cases, the CMA-ES significantly outpaces adaptive simulated annealing. For example, Fig. 2.2 shows the best points found by each algorithm when run 100 times with a fixed budget of 100, 300, 1,000 and 3,000 function evaluations. Clearly both algorithms are able to solve all three problems. But what takes ASA thousands of function calls, the CMA-ES achieves in a few hundred.

In this work, the CMA-ES is used almost exclusively. The following two subsections deal with explaining its inner workings in more complete detail. Initially, we will derive the basic update rules from an empirical point of view, emphasizing how to build up the CMA-ES from invariance principles and heuristics. We will also discuss the advantages this black-box optimizer affords over others. This practical, but unsystematic derivation for the CMA-ES may leave one wondering what exactly this algorithm has to do with evolution. Thus, the following section derives the same essential ingredients, but from the starting point of evolutionary game theory. This allows us a richer formalism to expand upon later, as well as elucidates some of the more intriguing properties connecting evolutionary computation to game theory and information geometry.

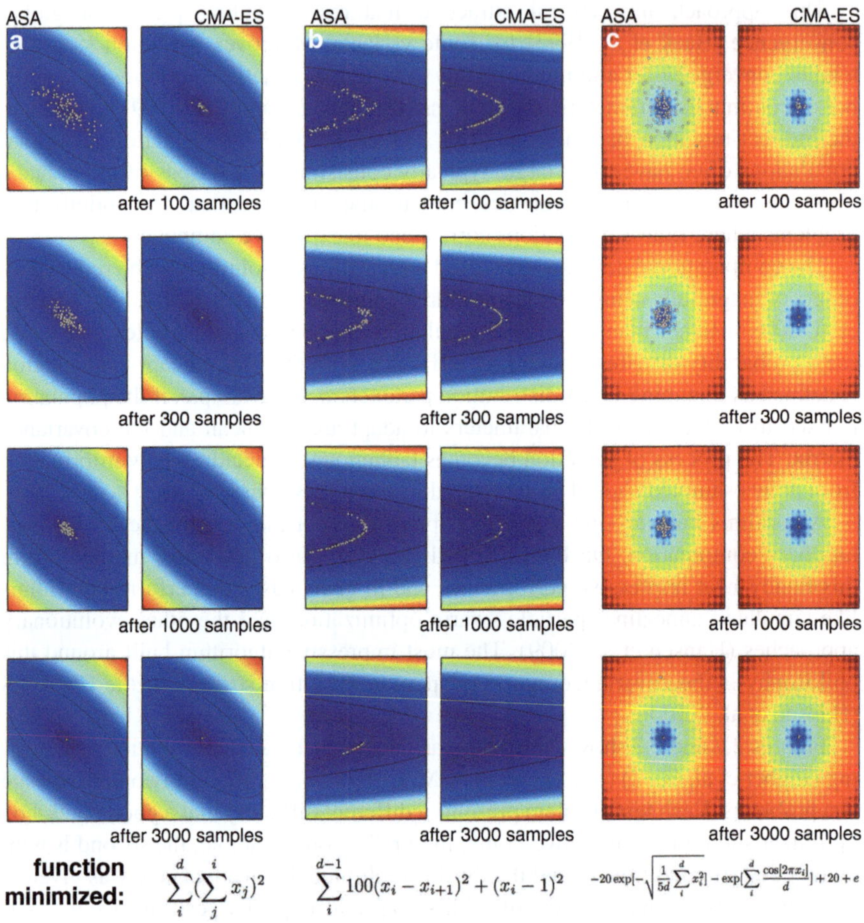

Fig. 2.2 A comparison between the CMA-ES and adaptive simulated annealing (ASA) on a non-seperable problem (**a**), an ill-conditioned problem (**b**), and a multimodal problem (**c**). The *white dots* in each panel correspond to the best object found during an independent run of each algorithm, after being given the number of function evaluations stated bellow. One can clearly see that in all three cases, both algorithms converge, but the CMA-ES does so significantly faster, typically completing the task after a few hundred evaluations

2.2 Deriving the CMA-ES Heuristically

Suppose that we are given a Gaussian function with mean μ_i and covariance matrix Σ_{ij} and tasked with finding the point x that minimizes some unknown function $g(x)$. As a first step, we sample N points using these parameters. Naturally, some points have lower values of $g(x)$ than others, and so selection should promote these points to appear more frequently in the subsequent distribution. Since both μ_i and Σ_{ij} are averages, one way to reflect this is by setting the new mean and covariance

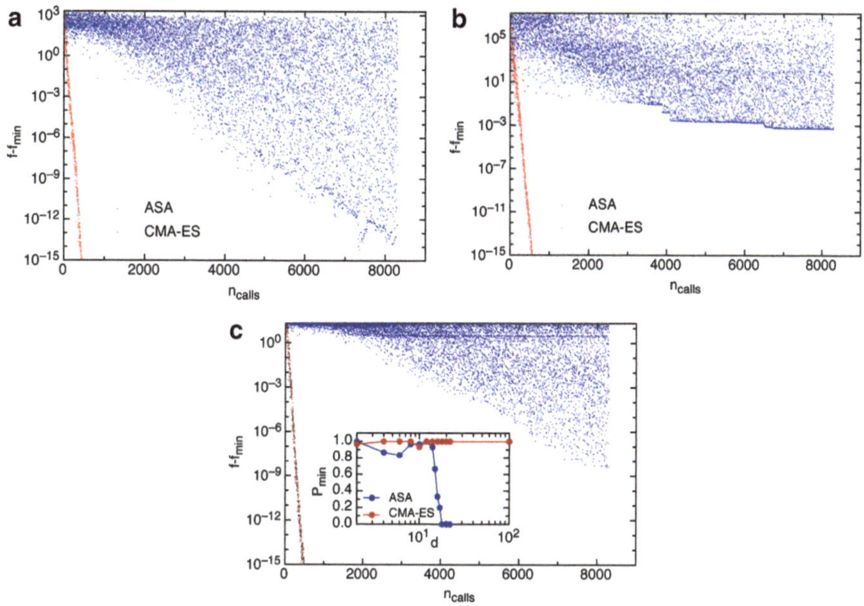

Fig. 2.3 Function values obtained against the number of evaluations made by the CMA-ES and ASA while solving the test problems shown in Fig. 2.2. Note problems (**a**) (**b**) and (**c**) correspond to the same labels from Fig. 2.2. While both algorithms achieve exponential convergence in terms of the best found solutions against time, the CMA-ES converges at a faster rate. In part, this is because of the fact that one updating cycle for the CMA-ES takes only six function evaluations whereas hundreds must be used by simulated annealing. Plotted in the inset for (**c**) is the probability of finding the global minimum for the d dimensional generalization of problem(c): $f = -20\exp[-0.2\sqrt{\frac{1}{d}\sum_i^d x_i^2}] - \exp[\frac{1}{d}\sum_i^d \cos[2\pi x_i]] + 20 + e$. At first, both algorithms scale to solve the problem in higher dimensions, with typical probabilities near 0.9. But ASA seems to fail in dimensions higher than 20, whereas the CMA-ES continue to find the global optimum even in a 100 dimensional search space without any parameter adjustment

matrix, μ_i' and Σ_{ij}', equal to weighted averages, emphasizing the best points in the generation. In other words, we can update the parameters as:

$$\mu_i' = \sum_x w(x) x_i \qquad \Sigma_{ij}' = \sum_x w(x)(x_i - \mu_i)(x_j - \mu_j) \tag{2.1}$$

$$\mu_i' = \mu_i + \sum_x w(x)(x_i - \mu_i) \quad \Sigma_{ij}' = \Sigma_{ij} + \sum_x w(x)[(x_i - \mu_i)(x_j - \mu_j) - \Sigma_{ij}] \tag{2.2}$$

where the last line follows assuming the weights sum to one. The transformation from the first to the second line both is to improve numerical stability and can be generalized to a moving average by multiplying the driving term with a small factor less than one.

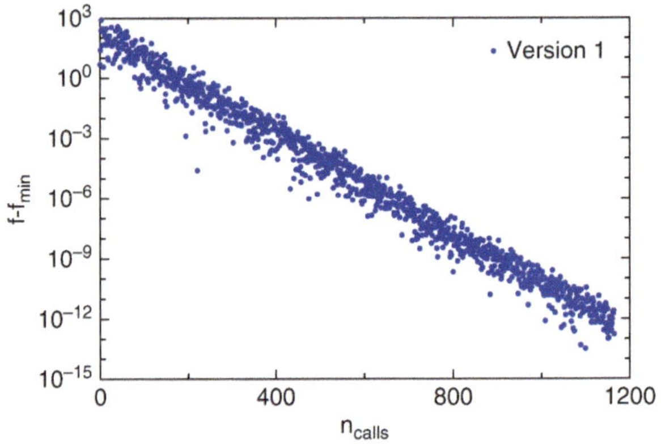

Fig. 2.4 Using Eqs. 2.1 to minimize the function $x_1^2 + x_2^2$. These simple update rules achieve a steady rate of exponential convergence on this and several other test problems

In addition, its very useful to pick fixed weights that only depend on the ranked position of $g(x)$. In other words, we pick a set of weights for all time-steps and match points to weights ignoring the magnitude of differences in $g(x)$ and focusing instead on their ordinal value. The logic here is two fold. First, an ideal optimizer should perform equally well on all rank preserving transformations of the original function because the positions of all the extrema are unchanged. Second, by using a rank based scheme, a steady pressure for improvement is applied, even if sampled values get arbitrarily close to the absolute function minimum.

Remarkably, these simple update rules are fairly good at solving optimization problems in and of themselves. In fact, Fig. 2.4 shows an algorithm using just 2.2 to minimize the parabolic function $x_1^2 + x_2^2$. The algorithm proves itself surprisingly efficient: it converges to the minima exponentially, reducing the function value roughly 1.3 orders of magnitude every 100 function evaluations.

Of course there are modifications that might improve the efficiency of this algorithm. After implementing a change, Fig. 2.2 will be revisited to illustrate the significance of each alteration. To keep track of the changes, each modification will be called by a different version number (note that these labels do not correspond to labels in the literature).

An obvious way to improve upon version 1 is to introduce a mechanism for amplification. For example, if every update to the mean points in the same direction, it makes sense to accelerate the rate the covariance matrix expands in that direction. This can be implemented by introducing a kind of damped momentum in the update

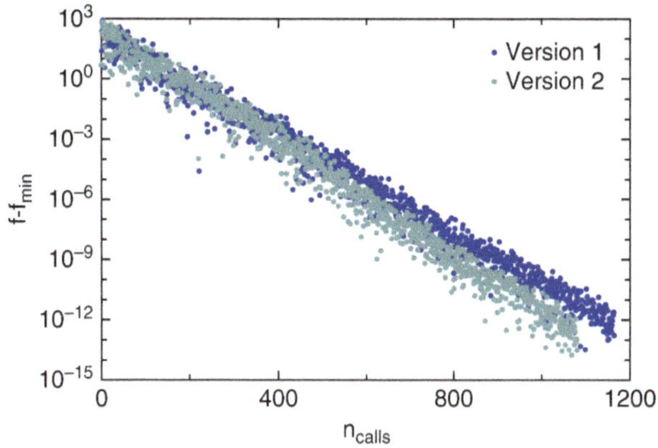

Fig. 2.5 Implementing a momentum addition via Eqs. 2.5 leads to a minor improvement over version 1. Specifically, the rate of convergence appears slightly faster. Thus, it would be difficult to observe any difference if the function were not minimized over several orders of magnitude

equations that stretches the covariance matrix along the typical direction the mean travels. Transforming these concepts into code produces version 2:

$$\mu_i' = \mu_i + \sum_x w(x)(x_i - \mu_i) \tag{2.3}$$

$$p_i\prime = (1 - c_1)p_i + c_2(\mu_i' - \mu_i) \tag{2.4}$$

$$\Sigma_{ij}' = (1 - c_3 - c_4)\Sigma_{ij} + c_3 \sum_x w(x)[(x_i - \mu_i)(x_j - \mu_j) - \Sigma_{ij}] + c_4 p_i' p_j' \tag{2.5}$$

where c_1, c_2, c_3 and c_4 are positive constants less than 1.

Figure 2.5 compares an implementation of version 1 against version 2, again minimizing the function $x_1^2 + x_2^2$. The updated version of the algorithm is technically faster, but not by much. We comment that the modifications to produce version 2 are included here to make a connection with the standard algorithm in the literature. On the basis of experience, however, we believe that updates using Eq. 2.5 are negligibly better than updates with Eq. 2.2, although appreciable more complicated. While these modifications are, as far as the literature goes, components of the complete CMA-ES algorithm, they do not seem to embody essential ones.

A more significant modification can be made by addressing the fact that the adaptations to the covariance matrix simultaneously change the shape and the size. It make sense to decouple these effects, in particular to produce amplification effects in scale when the covariance matrix is generally shrinking or expanding. To do this, one can split Σ_{ij} into two parts so that $\Sigma_{ij} = \sigma^2 C_{ij}$. Now, σ can scale Σ_{ij}, while C_{ij} determines its dominant directions.

The updates for σ can be developed by considering how far the mean actually moves relative to the width of the current covariance matrix. If the mean consistently moves less than one the standard deviation, it makes sense to start shrinking the width of Σ_{ij}. Conversely, if the mean consistently moves much further than the standard deviation, it makes sense to start scaling Σ_{ij} up. This heuristic can be encoded by introducing a momentum term q_i for σ and updating both as

$$q_i' = (1 - c_5)q_i + c_6\sqrt{\Sigma^{-1}}_{ij}(\mu_j' - \mu_j) \tag{2.6}$$

$$\sigma' = \sigma \exp[c_7(\frac{\|q_i'\|}{\langle\|N(0,I)\|\rangle} - 1)] \tag{2.7}$$

where the c terms are again small, positive constants and $\langle\|N(0,I)\|\rangle$ denotes the average length of a normally distributed vector of the same dimension. The idea here is that the vector q_i represents a weighted average of $\sqrt{\Sigma^{-1}}_{ij}(\mu_j' - \mu_j)$ from each step. If most of these moves fall within the standard deviation, then $(\mu_i' - \mu_i)\Sigma_{ij}^{-1}(\mu_j' - \mu_j) \approx \langle\|N(0,I)\|^2\rangle$. When this happens, the update rule for σ yields no change, while smaller deviations lead to contraction and larger deviations lead to expansion.

Separating Σ_{ij} into the scale and direction components also requires the removal of any bias in changes to C_{ij} from adjusting σ independently. This is achieved by changing rules (2.4) and (2.5) into

$$p_{i'} = (1 - c_1)p_i + c_2\frac{\mu_i' - \mu_i}{\sigma'} \tag{2.8}$$

$$C_{ij}' = (1 - c_3 - c_4)C_{ij} + c_3\sum_x w(x)[\frac{x_i - \mu_i}{\sigma'}\frac{x_j - \mu_j}{\sigma'} - C_{ij}] + c_4 p_i' p_j' \tag{2.9}$$

As seen in Fig. 2.6, separating the covariance matrix into shape and scale parameters yields a significant payoff in speed. Indeed, the rate of convergence has been enhanced from 1.3 orders of magnitude per hundred function evaluations to 3. Practically speaking, this means that in the amount of time it takes version 1 to finish a single optimization, version 3 can be run almost three times.

The full algorithm is the complete CMA-ES. For the sake of clarity, it has been reproduced bellow in psuedocode.

2.2.1 Invariance Properties

The justifications used at each step to produce the CMA-ES are extremely heuristic. Thus, it begs the question, why exactly does this algorithm work so well? A popular speculation is that the CMA-ES boasts a large number of invariance properties (Hansen 2006; Ollivier et al. 2011). By invariance, we mean the algorithm's performance will not be significantly impacted by a given problem transformation. These invariances include:

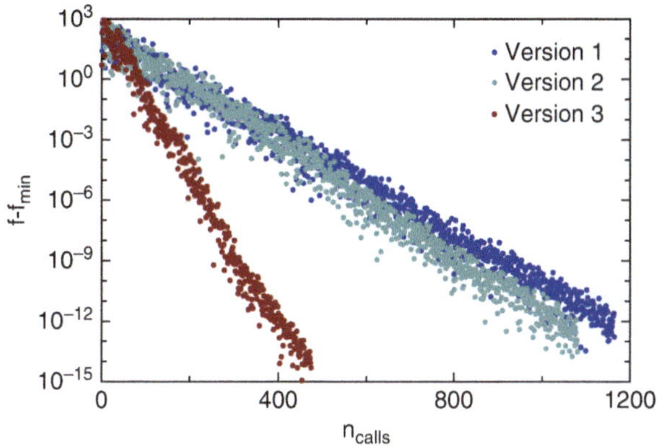

Fig. 2.6 Decoupling the width of the covariance matrix from its shape produces a significant decrease in the number of function calls needed to complete an optimization. Specifically, the rate of convergence is enhanced by almost a factor of 3

Algorithm 1 the CMA-ES

Require: c, w, σ, μ, population size

$p_i \leftarrow 0$

$q_i \leftarrow 0$

$C_{ij} \leftarrow \delta_{ij}$

while g(x) $\geq g_{tol}$ **do**

Get a population of samples $x \leftarrow N(\mu, \sigma^2 C)$

Produce the weighted list w(x) for the samples given each value of g(x)

$\mu'_i \leftarrow \mu_i + \sum_x w(x)(x_i - \mu_i)$

$q'_i \leftarrow (1 - c_5)q_i + c_6\sqrt{\Sigma^{-1}}_{ij}(\mu'_j - \mu_j)$

$p'_i \leftarrow (1 - c_1)p_i + c_2\frac{\mu'_i - \mu_i}{\sigma'}$

$C'_{ij} \leftarrow (1 - c_3 - c_4)C_{ij} + c_3\sum_x w(x)[\frac{x_i - \mu_i}{\sigma'}\frac{x_j - \mu_j}{\sigma'} - C_{ij}] + c_4 p'_i p'_j$

$\sigma' \leftarrow \sigma \exp[c_7(\frac{\|q'_i\|}{\langle\|N(0,I)\|\rangle} - 1)]$

end while

- Invariance to translations: clearly any serious optimizer must have this property, but explicitly the mean μ_i is a learned parameter. Therefore, translating the search space so that $x_i \to x_i + a_i$, where a_i is a constant, should not seriously impact the algorithm performance.
- Invariance to rotations: since the CMA-ES learns the full covariance matrix, it is able to adapt to rotations in the input parameter space. That is, the algorithm is invariant under $x_i \to R_{ij}x_j$ where R_{ij} is a rotation matrix. This feature is extremely useful since makes it irrelevant as to whether or not the natural parameters for a problem are linear combinations of the parameters given to the optimizer.
- Invariance to scaling: again, owing to the fact that the CMA-ES learns the full covariance matrix automatically, the algorithm will perform equally well on problems where $x_i \to bx_i$ where b is a positive scale factor.

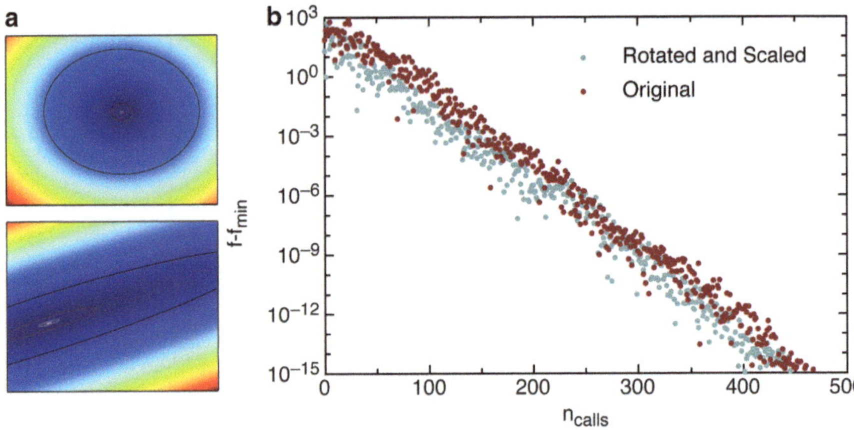

Fig. 2.7 (**a**) A parabolic test function and the same function after a random rotation and scaling. If the CMA-ES is run on each of these problems, the performance of the algorithm, as characterized by the fitness vs. time graph is practically unchanged

- Invariance to monotone fitness transforms: Because the algorithm updates itself only using ranks of the sampled points, CMA-ES behaves identically when solving any order preserving transformation of the original problem. That is, the CMA-ES is invariant under $g(x) \to f(g(x))$ provided $f(x)$ leaves the ranks intact. Examples of this transformation include constant scaling, offsets, or even exponentiation.

To demonstrate the utility of these invariances, Fig. 2.7 compares two independent runs of the CMA-ES started with the same initial mean and covariance matrix. One run optimizes a simple parabolic function while the other optimizes a randomly translated, rotated, and scaled version of the same problem. In spite of these distortions, the fitness values found at each step are essentially identical, asserting that if the CMA-ES is run on two versions of the same problem, differing only by an affine transformation, the performance will be indistinguishable.

Finally, its worth noting that, while there are a number of free parameters c, in practice they rarely have a significant impact on algorithm performance. In fact, they can be typically be set using empirical relations that depend only on the problem dimension (Hansen et al. 2003; Hansen 2006). The result is that, from a user perspective, the CMA-ES is parameter free. Practically speaking this is a major benefit since it means that few, if any, optimization runs have to be spent on calibration. Instead, the optimizer can be put to work immediately on problem solving.

2.2.2 The Benefit of Parameterization

There can be something dissatisfying about parameterizing the search distribution as a Gaussian. It seems it would be better to allow the algorithm to pick a distribution that best reflects the global search landscape. Indeed, simulated annealing, genetic algorithms and traditional evolution strategies are all built around the philosophy of particle based, implicit distributions. Yet all three of these algorithms are typically outperformed by the CMA-ES.

We speculate that part of the reason for this incongruity is that implicit distributions demand a tremendous amount of information to be statistically sound, and yet most of this information is inevitably discarded by the optimizer. For example, simulated annealing typically needs to take a large (say order 100) number of steps at each iteration to ensure that the sampling is in fact from a Gibbs distribution with the prescribed temperature. Thus computational effort is, in part, devoted to simply generating new sample points. Conversely, in parameterized distributions, points are sampled with the sole purpose of updating parameters. This means that, in principle, one only needs to generate as many points as there are open parameters, yielding a small and systematized demand.

To put this in perspective, consider how much cost is expended to produce any change in the average solution quality. For simulated annealing, the average fitness changes every time a new step is reached in the annealing schedule, (i.e. every time the temperature changes), and to achieve this goal, typically a few hundred function calls have to be expended. The same is true for a population based evolutionary algorithm: to go from one generation to the next, a few hundred candidate solutions need to be evaluated. Both cases stand in stark contrast to the CMA-ES where an average change in fitness only costs a few dozen function evaluations.

Because randomized algorithms are developed by considering how expectation values or distributions behave, it is easy to ignore the demands and difficulties associated with constructing these objects. If the sampling method is elaborate, needs to converge, or if the population is implicit, a large number of function evaluations can be needed only to make the logic behind the algorithm statistically sound. Conversely, if the samples are generated exactly from a prescribed distribution, several sources of error are eliminated a priori, allowing statistically similar improvements to be achieved at a lower computational cost.

The potential disadvantage of parameterizing the search distribution is that it can only examine one specific region, thereby excluding the possibility of simultaneously exploring multiple, distant minima. But given this concern, its worth asking how many function evaluations are needed before a particle based method gains the advantage. For example, a simulated annealing algorithm trapped in a local well should always find its way to the global minima, provided it is given infinitely many function evaluations (Kirkpatrick et al. 1983). But in practice, it can become the case that the algorithm is not given enough time to leave the well, thereby reducing it to an effectively local search method. More generally, both simulated annealing and types of evolution strategies are guaranteed to find global optima on search problems given an infinitely large number of evaluations per iteration (Back et al. 1993; Kirkpatrick et al. 1983). But in practice, realizing the conditions for these proofs on complicated problems can require demands beyond either the expectations or inclinations of a user. When this happens, it becomes more practical to converge quickly with a local search algorithm and restart than to try and distinguish the global optimum in one fell swoop.

Altogether these results emphasize the tradeoff between algorithm generality and algorithm speed. At a theoretical level, population based methods favor the former while parameterized methods, like the CMA-ES, favor the later. Yet in practice, speed is typically the deciding factor for whether an algorithm is useful. To the extent that this is true, parameterized algorithms gain a measurable advantage over particle based algorithms, as reflected in empirical benchmarking (Hansen et al. 2009).

2.3 Deriving Evolutionary Algorithms from Evolutionary Game Theory

At first glance its hard to see what, if anything, the CMA-ES has to do with evolution in the biological sense of the word. The heuristic derivation plus the parameterized distribution seem to have little to do with the concepts of selection or survival of the fittest. Yet in this section we will show that algorithms in the same class as the CMA-ES share a deep connection with evolutionary game theory. In fact, using evolutionary game theory as a jumping off point one can reproduce several of the main ideas that make the CMA-ES so powerful.

The objective here is not to completely reproduce the CMA-ES, but rather to find a framework that allows its key features to be generalized. To the extent that the CMA-ES is successful because of invariance properties, the core of the algorithm is defined by the original weighted mean and covariance matrix updates in Eq. 2.1, and the rank-based weighting scheme. Thus the goal becomes to derive these features, which can be nicely met by starting with the fundamental equation in evolutionary game theory (Hofbauer and Sigmund 2003).

2.3.1 The Replicator Equation

Consider the simple model for a biological system where objects reproduce by replication (Hofbauer and Sigmund 2003). That is, each object makes identical copies of itself. Further, the rate at which objects replicate is set by the quality or fitness of an object x_i, denoted by $f(x_i)$:

$$\dot{n}_i = n_i f(x_i) \tag{2.10}$$

where n_i is the number of x_i objects. Moreover, the probability that a randomly selected object is of the type x_i is $P_i = n_i / \Sigma_i n_i$ and, given (2.10) this probability evolves as

$$\dot{P}_i = \frac{\dot{n}_i}{\sum_i n_i} - \frac{n_i}{\sum_i n_i} \frac{\sum_i \dot{n}_i}{\sum_i n_i} \tag{2.11}$$

$$\dot{P}_i = \frac{n_i f(x_i)}{\sum_i n_i} - \frac{n_i}{\sum_i n_i} \frac{\sum_i n_i f(x_i)}{\sum_i n_i} \tag{2.12}$$

$$\dot{P}_i = P_i(f(x_i) - \langle f \rangle) \tag{2.13}$$

where the angle brackets denote averaging over all possible objects x_i. The system can be generalized to a continuous framework by replacing n_i with the number density $n(x)$ and the probability P_i with the probability density $\rho(x)$, or

$$\dot{\rho}(x) = \rho(x)[f(x) - \langle f \rangle] \tag{2.14}$$

This equation describes some of the basic concept of evolution, but its not complete. Clearly, objects fitter than average proliferate while the probability of selecting objects that are less fit than the average decays to zero. However, as it stands objects don't compete, they only reproduce. It would be more appropriate if increasing the number of fit objects directly caused the reproduction rate $f(x)$ for unfit objects to suffer as well. This can be accomplished by introducing a payoff function $\Pi(x, x')$ and setting $f(x) = \int dx' \Pi(x, x') \rho(x')$. This payoff function specifies the amount of reward an object at x gets when competing against an object at x'. The fitness is then defined as the average payoff when competing against members of the population drawn at random. For example, if we choose $\Pi(x, x') = \mathbb{1}[g(x') > g(x)]$ so that an object at point x is awarded one unit of fitness every time it encounters an object with a larger g, then averaged over the whole population, the fitness $f(x)$ equals the probability of having a lower value of g than another object drawn randomly from the population. Written explicitly, a payoff function changes Eq. 2.14 into

$$\dot{\rho}(x) = \rho(x)[\int dx' \Pi(x, x') \rho(x') - \int \int dx dx' \Pi(x, x') \rho(x') \rho(x)] \tag{2.15}$$

although in what follows, we will often keep $f(x)$ as shorthand.

Equation 2.15 is known as the continuous replicator equation (Oechssler and Riedel 2002, 2001; Cressman 2005) and appears in diverse fields ranging from game theory to ecology to economics (Hofbauer and Sigmund 2003). In the following sections, we will show how some special manipulations can also connect it with the CMA-ES.

2.3.2 Parameterizing the Replicator Equation

At a glance, its hard to see how exactly the replicator equation can be used as an algorithm. To evaluate the replicator equation as is, one would need to calculate the fitness at every point. This makes the equation irrelevant for design and optimization since evaluating every point is tantamount to brute forcing the problem.

However, the continuous replicator equation can be turned into something more practical by constraining the distribution to have a fixed structure. For example, the distribution ρ could be fixed to always be a Gaussian or a Cauchy functional form. Doing so makes ρ a function of both x and a finite number of parameters λ_i. Continuing with the Gaussian example, the parameters λ could be the mean μ and the covariance matrix Σ (though other choices are possible). With the form of the distribution fixed, these parameters become the sole source of time dependance. This transforms Eq. 2.14 into

$$\dot{\rho}(x|\lambda) = \rho(x|\lambda)[f(x) - \langle f \rangle] \tag{2.16}$$

$$\frac{\partial \log(\rho)}{\partial \lambda_i}\dot{\lambda}_i = [f(x) - \langle f(x) \rangle] \tag{2.17}$$

Now, the trick is to isolate a set of update rules for each λ_i that depend only on averages, rather than point-wise values. Once this is done, averages can be approximated by sampling from the distribution, thereby avoiding the issue of analyzing every point. This goal is met with a bit of algebra. First we multiply both sides by $\rho\frac{\partial \log(\rho)}{\partial \lambda_j}$ and integrate to get

$$\langle \frac{\partial \log(\rho)}{\partial \lambda_i}\frac{\partial \log(\rho)}{\partial \lambda_j} \rangle \dot{\lambda}_i = \langle \frac{\partial \log(\rho)}{\partial \lambda_j}[f(x) - \langle f(x) \rangle] \rangle \tag{2.18}$$

The matrix on the right hand side can then be inverted to produce

$$\dot{\lambda}_i = \langle \frac{\partial \log(\rho)}{\partial \lambda_i}\frac{\partial \log(\rho)}{\partial \lambda_j} \rangle^{-1} \langle \frac{\partial \log(\rho)}{\partial \lambda_j}[f(x) - \langle f(x) \rangle] \rangle \tag{2.19}$$

This equation is also has a name: the information geometric optimization flow (Wierstra et al. 2008; Ollivier et al. 2011). In recent work, it has been proposed as a unifying framework for black-box optimizers in the same class as the CMA-ES (Akimoto et al. 2010; Ollivier et al. 2011), although prior work has not yet connected it to the replicator equation. Instead, it has been derived abstractly with concepts from information theory and differential geometry. But, before examining the ramifications of connecting these two disciplines, we turn to examining in greater detail the content behind Eq. 2.19.

2.3.3 Evolution Equations for the Exponential Family

To see how exactly Eq. 2.19 works, one can pick a family of distributions and examine the consequences. A natural choice is the exponential family in which distributions are of the form $\exp[T_i(x)\lambda_i]/\int dx \exp[T_i(x)\lambda_i]$. Because exponential family distributions have the property that $\partial_{\lambda_i}\log[\rho] = T_i(x) - \langle T_i(x)\rangle$, the resulting update equations simplify significantly:

$$\dot{\lambda}_i = \langle \partial_{\lambda_i}\log[\rho]\partial_{\lambda_j}\log[\rho]\rangle^{-1}\langle \partial_{\lambda_j}\log[\rho]f(x)\rangle \tag{2.20}$$

$$\dot{\lambda}_i = \langle (T_i - \langle T_i\rangle)(T_j - \langle T_j\rangle)\rangle^{-1}\langle (T_j - \langle T_j\rangle)f(x)\rangle \tag{2.21}$$

$$\dot{\lambda}_i = Cov[T_i, T_j]^{-1}Cov[T_j, f(x)] \tag{2.22}$$

Apparently, the parameters λ_i adapt themselves to follow the correlations between T_i and $f(x)$, although the exact process is obscured by the matrix inversion. A clearer perspective can be obtained by examining how the averages $\langle T_i\rangle$ evolve

$$\frac{d}{dt}\langle T_i\rangle = \partial_{\lambda_k}\langle T_i\rangle\dot{\lambda}_k = \langle T_i\partial_{\lambda_k}\log[\rho]\rangle\langle \partial_{\lambda_k}\log[\rho]\partial_{\lambda_j}\log[\rho]\rangle^{-1}\langle \partial_{\lambda_j}\log[\rho](f - \langle f\rangle)\rangle \tag{2.23}$$

Replacing all of the $\partial_{\lambda_i}\log[\rho]$ terms by $T_i(x) - \langle T_i(x)\rangle$ gives

$$\frac{d}{dt}\langle T_i\rangle = \langle T_i(T_k - \langle T_k\rangle)\rangle\langle (T_k - \langle T_k\rangle(T_j - \langle T_j\rangle)\rangle^{-1}\langle (T_j(x) - \langle T_j(x)\rangle(f - \langle f\rangle)\rangle \tag{2.24}$$

$$\frac{d}{dt}\langle T_i\rangle = Cov[T_i, T_k]Cov[T_k, T_j]^{-1}Cov[T_j, f] \tag{2.25}$$

$$\frac{d}{dt}\langle T_i\rangle = Cov[T_i, f] \tag{2.26}$$

Very simply, we find that the averages conjugate to each λ_i evolve to directly follow correlations in fitness. This provides a clearcut interpretation. For example, if we take $T_i = x_i$ then updating with Eq. 2.19 creates a current of probability flux pointing in the direction most correlated with improvements in solution quality: $\frac{d}{dt}\langle x_i\rangle = Cov[x_i, f]$.

2.3.4 Evolution for a Gaussian Distribution

If the search distribution is taken to be a multivariate Gaussian, the results in the prior section can be used to derive its update rules. Specifically, define $\lambda = [\Sigma_{ij}^{-1}\mu_j, -\frac{1}{2}\Sigma_{ij}^{-1}]$ so that $T = [x_i, x_ix_j]$. Inserting this into the update equation for the moments, (2.26), gives

$$\frac{d}{dt}\langle x_i\rangle = Cov[x_i,f] \qquad \frac{d}{dt}\langle x_ix_j\rangle = Cov[x_ix_j,f] \tag{2.27}$$

This result can be rewritten in terms of the mean and covariance matrix only:

$$\frac{d}{dt}\mu_i = Cov[x_i,f] \qquad \frac{d}{dt}\Sigma_{ij} = Cov[(x_i-\mu_i)(x_j-\mu_j),f] \tag{2.28}$$

Structurally, Eq. 2.28 is the differential equation version of the CMA-ES Eq. 2.1.

A potential difference is the presence of a rank-based weighting function in Eq. 2.1 versus the fitness function, $f(x)$, in Eq. 2.28. But recall that, for a replicator equation with selection, $f(x) = \int dx'\,\Pi(x,x')\rho(x')$. By inserting a payoff function, the correspondence can be completed. For example, the previous choice $\Pi(x,x') = \mathbb{1}[g(x') > g(x)]$ produces weights that depend only on the ranked values of each object. Specifically, this payoff function is equivalent to the weighting scheme $w(x) \propto P(x)$ where P is the probability that the fitness at x beats another object drawn randomly from the distribution. Generally, weighting schemes can be matched to payoff functions by the relation $w(x) \propto \int dx'\,\Pi(x,x')\rho(x')$.

At this point, the goal of generalizing all the key invariances boasted by the CMA-ES has been fulfilled. Moreover, the presence of a clear correspondence between biological evolution and the CMA-ES suggest a connection with natural evolution is less tenuous than it first seems: the core concepts that the CMA-ES is built around can be linked to fundamental elements in biology. Explicitly, the CMA-ES as originally defined by Eqs. 2.1 can be thought of a parameterized model for replicating and competing species, in which selection pressure drives changes to the distribution.

2.3.5 *Solving the Replicator Equation with Exponential Families*

Putting optimization aside for the moment, the correspondence between evolutionary algorithms and the replicator equation can also be used for more theoretical ends. As it turns out, exponential families are also a good ansatz for creating exact solutions to the full replicator equation 2.15. As a general example, consider an infinite exponential family of the form $\exp[\Sigma_n\lambda_n x^n]$ (for ease of notation, the x^n is being used symbolically and the sum is carried out over all order n combinations of the x variables: $x^n = x_ix_jx_k\ldots$). In this case, the evolution equations for the average values $\langle x^n\rangle$ are

$$\frac{d}{dt}\langle x^n\rangle = Cov[x^n,f] \tag{2.29}$$

But calculating the evolution equation for the moments from the full replicator equation produces

$$\frac{d}{dt}\langle x^n\rangle = \int dx\dot{\rho}x^n = \langle \rho x^n(f - \langle f\rangle)\rangle = Cov[x^n, f] \tag{2.30}$$

Evidently, all the moments evolve identically in the parameterized infinite exponential family model and the original replicator equation, implying the two distributions are indistinguishable.

Whats more, the power series solution implied by an infinite family can be reduced if the payoff function is of a special form. Specifically, if the payoff function $\Pi(x, x')$ can be factored into $\Pi(x, x') = \sum_n \Psi_n(x)\Phi_n(x')$, then a distribution of the form $\exp[\sum_n \Psi_n(x)\lambda_n(t)]$ is left invariant under the replicator equation, provided $\dot{\lambda}_n(t) = \langle\Phi_n\rangle$. Furthermore, the parameterized solution for the corresponding evolutionary algorithm yields exactly these dynamics. The proof is given by direct substitution into (2.15):

$$\dot{\rho}(x) = \rho(x)[\int dx'\Pi(x, x')\rho(x') - \int\int dxdx'\Pi(x, x')\rho(x')\rho(x)] \tag{2.31}$$

$$\frac{d}{dt}\log[\rho(x)] = [\int dx'\Pi(x, x')\rho(x') - \int\int dxdx'\Pi(x, x')\rho(x')\rho(x)] \tag{2.32}$$

inserting $\Pi(x, x') = \Phi_n(x')\Psi_n(x)$ and $\rho = \exp[\sum_n \Psi_n(x)\lambda_n(t)] / \int dx \exp[\sum_n \Psi_n(x)\lambda_n(t)]$ we get

$$\sum_n[\Psi_n(x)\dot{\lambda}_n(t) - \langle\Psi_n\rangle\dot{\lambda}_n(t)] = \sum_n[\Psi_n(x)\langle\Phi_n\rangle - \langle\Psi_n\rangle\langle\Phi_n\rangle] \tag{2.33}$$

Equality is obtained when $\dot{\lambda}_n = \langle\Phi_n\rangle$.

This equality is produced by employing the corresponding evolutionary algorithm. In this case, the coupling constants λ_n are governed by Eq. 2.22. Inserting $\Psi_n(x)$ as the conjugate function to each λ_n gives

$$\dot{\lambda}_n(t) = Cov[\Psi_n, \Psi_m]^{-1}Cov[\Psi_m, f(x)] \tag{2.34}$$

But by inserting the assumed for the payoff function we find

$$Cov[\Psi_m, f] = Cov[\Psi_m, \Psi_n\langle\Phi_n\rangle] \tag{2.35}$$

producing the final equation for the coupling constants

$$\dot{\lambda}_n(t) = Cov[\Psi_n, \Psi_m]^{-1}Cov[\Psi_m, f] \tag{2.36}$$

$$\dot{\lambda}_n(t) = Cov[\Psi_n, \Psi_m]^{-1}Cov[\Psi_m, \Psi_k]\langle\Phi_k\rangle \tag{2.37}$$

$$\dot{\lambda}_n(t) = \langle\Phi_n\rangle \tag{2.38}$$

which is the requirement to solve the full replicator equation. Altogether we find that if the payoff function can be factored, there is an evolutionary algorithm that creates exact solutions to the full replicator equation.

On the one hand, this discussion suggests that, at some level, evolutionary algorithms are a formalism for solving classes of replicator equations. On the other hand, it is an active subject of research in evolutionary computation on how to best choose weighting functions (Wierstra et al. 2008). This analysis suggests that the payoff functions, and by extension weighting schemes, might best be selected by minimizing the difference between the true replicator equation's evolution and the parameterized one. Evidently, putting evolutionary algorithms in contact with theoretical biology invites new perspectives for natural and artificial evolution alike.

2.3.6 Global Convergence for Replicator Equations

At present, we have a systematic framework to derive new algorithms with the parameterized replicator equation 2.19. But so far little has been said to about any practical consequences of starting from the replicator equation. One potential benefit is that prior work has yielded conditions under which replicator equations converge to the global optimum (Oechssler and Riedel 2002; Cressman 2005). Thus it is hopeful that evolutionary algorithms built from the same concept might inherent this feature.

By way of this point, consider the following sketch of convergence. We start by considering what happens to the probability of not drawing a sample in a given region, $K = 1 - P_A$ where P_A is the probability of drawing a sample that lives in region A. For the time being, suppose A is a small sphere around an arbitrary test point. If the density evolves according to the replicator equation, then we find that K changes via

$$\frac{d}{dt}K = -\frac{d}{dt}P_A = -\langle (f - \langle f \rangle)\mathbb{1}_A \rangle \tag{2.39}$$

where $\mathbb{1}_A$ is 1 whenever the sampled point is in the region defined by A and 0 otherwise. Assume that f has a special point x_{opt} where the fitness is always greater than or equal to the population average. Then, by picking A as the region infinitesimally surrounding x_{opt}, we find $\frac{d}{dt}K < 0$ when $P_A < 1$. Further, $K > 0$ as long as $P_A < 1$. Combined, these two facts mean that K is a Lyapunov function and $P_A \rightarrow 1$ as the system is left to evolve.

The result here may not be rigorous, but its easy to rationalize: the point that maximizes the fitness always increases in density, so in the long time limit, all the density must wind up here. The skeptical reader is directed to the complicated but exact proofs of convergence in the literature (Oechssler and Riedel 2002; Cressman 2005).

In contrast to the replicator equation, rigorous proofs explaining the convergence of evolutionary algorithms like the CMA-ES have remained elusive for over 10 years. One route to this end might be to compare the dissimilarity between an evolutionary optimizer's probability distribution and the exact replication equation solution it is approximating. If one can demonstrate that the approximation converges to the exact equation in the long time limit, then it must also inherit the global convergence properties. To our knowledge this approach is unexplored and seems like a promising route for explaining and expanding evolutionary computation's current state of the art.

2.3.7 *Change in Entropy*

By definition evolution is a process that exchanges diversity and quality. To make an optimizer, we want to transform the former to the later, but in a balanced way. If we trade the range of potential solutions for improvements too quickly, then its likely the optimizer will get stuck in a local optimum. Conversely, in order to converge, the optimizer must give up some diversity in each iteration.

To analyze this exchange, we define the change of diversity of a population in terms of its change in relative entropy (Ollivier et al. 2011) . For two distributions, $\rho(t)$ and $\rho(t+\delta t)$ that are infinitesimally close to one another, one measure for the loss of diversity is the relative change in entropy or $\int \rho(t)\log(\rho(t)) - \int \rho(t)\log(\rho(t+\delta t))$. Taylor expanding to second order in δt gives

$$\delta S = \int \rho(t)\log[\rho(t)] - \int \rho(t)\log[\rho(t+\delta t)] = -\int \partial_t \rho \delta t - \int \rho \partial_{tt}\log[\rho]\delta t^2 \tag{2.40}$$

$$\delta S = -\delta t^2 \int \rho \ddot{\lambda}_i \partial_{\lambda_i}\log[\rho] - \delta t^2 \int \rho \dot{\lambda}_i \dot{\lambda}_j \partial_{\lambda_i \lambda_j}\log[\rho] \tag{2.41}$$

$$\delta S = \delta t^2 \dot{\lambda}_i \dot{\lambda}_j \int \rho \partial_{\lambda_i}\log[\rho]\partial_{\lambda_j}\log[\rho] \tag{2.42}$$

$$\delta S = \delta t^2 \dot{\lambda}_i \dot{\lambda}_j \langle \partial_{\lambda_i}\log[\rho]\partial_{\lambda_j}\log[\rho]\rangle \tag{2.43}$$

Since the first order change in relative entropy as defined above always vanishes, this term represents the leading order increment along any path the parameters might take. If we follow the specific path of (2.19), then replacing one of the $\dot{\lambda}_i$ terms gives

$$\delta S = \delta t^2 \dot{\lambda}_i \langle \partial_{\lambda_j}\log[\rho]\partial_{\lambda_k}\log[\rho]\rangle^{-1}\langle \partial_{\lambda_i}\log[\rho]\partial_{\lambda_j}\log[\rho]\rangle\langle \partial_{\lambda_k}\log[\rho](f - \langle f\rangle)\rangle \tag{2.44}$$

$$\delta S = \delta t^2 \dot{\lambda}_i \langle \partial_{\lambda_i}\log[\rho](f - \langle f\rangle)\rangle \tag{2.45}$$

$$\delta S = \delta t^2 \langle \partial_t \log[\rho](f - \langle f\rangle)\rangle \tag{2.46}$$

$$\delta S = \delta t^2 \partial_t \langle f - \langle f \rangle \rangle - \langle \partial_t (f - \langle f \rangle) \rangle \tag{2.47}$$

$$\delta S = -\delta_t{}^2 \langle \partial_t (f - \langle f \rangle) \rangle \tag{2.48}$$

Eq. (2.48) has a nice physical interpretation associated with it. By rearranging the terms slightly it states

$$\frac{\delta S}{\delta t} = -\langle \delta (f - \langle f \rangle) \rangle \tag{2.49}$$

In other words, the rate that entropy (i.e. diversity) changes over a small, finite step is proportional to the population averaged change in fitness. In this sense, (2.19) makes an explicit statement about how diversity loss and fitness improvement are intertwined. Selecting a particular fitness function f is equivalent to deciding how fast diversity should be lost when the fitness increases by a fixed amount. Colloquially, if one is able to produce an average increase in fitness of q then one is willing to loose diversity at a rate of q.

2.4 Computation Time and Evolution

In an ideal world, the ability of an optimizer to solve a given problem would only depend upon the how rugged, discontinuous, or misshapen the minimization task is. In the real world, the dominant constraint is sufficient computer power. Even if the input-output relationship used by the optimizer is simple, optimizers can still be ineffective if the time to calculate an output is excessively long.

In light of this, we estimate that evolutionary optimization only recently became a general method for materials design. Between 1986 and 2007, Hilbert et al. (Hilbert and Lopez 2011) estimate that general purpose computing power doubled roughly every 18 months. This implies that a currently feasible simulation based optimization method for materials would have been impossibly slow 5 years ago, and will be extremely fast if implemented 5 years from now. As a specific example, consider the simulations in the following chapter. When run on a computer purchased in 2008, it took roughly 600 s to complete one simulation while the full optimization took roughly half a month to complete. Consistent with the rate of increase, the same optimization can be completed in a few days using hardware purchased in 2013, with simulations finishing every 250 s. Likewise, without specialized hardware, the same optimization would have taken roughly of 3 years to complete in 2000.

Taken optimistically, this result also implies that an automated materials design framework could transform exponential growth in computer power into exponential growth in optimized materials. To the extent that materials properties are collective phenomena, there is a lower bound of compute power needed to be near the thermodynamic limit for a given simulation model. Furthermore, adding elements beyond this boundary in the same model may not substantially alter predictions about macroscopic properties. To the extent that this holds true, an increase in computer power every 18 months would implies that the time to go from concept to prototype halves every 2 years in automated materials design, while the scope of designable materials doubles at the same rate.

References

Akimoto, Y., Nagata, Y., Ono, I., & Kobayashi, S. (2010). Bidirectional relation between CMA evolution strategies and natural evolution strategies. In *Parallel problem solving from nature, PPSN XI* (pp. 154–163). Berlin/Heidelberg: Springer.

Back, T., Hoffmeister, F., & Schwefel, H. P. (1991). A survey of evolution strategies. In *Proceedings of the 4th International Conference on Genetic Algorithms*, San Diego, San Francisco (pp. 2–9).

Back, T., Rudolph, G., & Schwefel, H.-P. (1993). Evolutionary programming and evolution strategies: Similarities and differences. In *Proceedings of the Second Annual Conference on Evolutionary Programming*, San Diego, La Jolla (pp. 11–22).

Beyer, H. G., & Schwefel, H. P. (2002). Evolution strategies – A comprehensive introduction. *Natural Computing, 1*(1), 3–52.

Cressman, R. (2005). Stability of the replicator equation with continuous strategy space. *Mathematical Social Sciences, 50*(2), 127–147.

de Jong, K. (1975). An analysis of the behavior of a class of genetic adaptive systems. Ph. D. thesis, University of Michigan.

Eiben, A. E., & Smith, J. E. (2003). *Introduction to evolutionary computing*. New York: Springer.

Hansen, N. (2006). The CMA evolution strategy: A comparing review. In *Towards a new evolutionary computation* (pp. 75–102). Berlin/Heidelberg: Springer.

Hansen, N., Auger, A., Ros, R., Finck, S., & Posik, P. (2009). Comparing results of 31 algorithms from the black-box optimization benchmarking bbob-2009. In *Proceedings of the 12th Annual Conference Companion on Genetic and Evolutionary Computation* (pp. 1689–1696). New York: ACM.

Hansen, N., Muller, S. D., & Koumoutsakos, P. (2003). Reducing the time complexity of the derandomized evolution strategy with covariance matrix adaptation (CMAES). *Evolutionary Computation, 11*(1), 1–18.

Hilbert, M., & Lopez, P. (2011). The world's technological capacity to store, communicate, and compute information. *Science, 332*(6025), 60–65.

Hofbauer, J., & Sigmund, K. (2003). Evolutionary game dynamics. *Bulletin of the American Mathematical Society, 40*(4), 479–519.

Hoffmeister, F., & Back, T. (1991). Genetic algorithms and evolution strategies: Similarities and differences. In *Parallel problem solving from nature*. Berlin/Heidelberg: Springer.

Holland, J. H. (1973). Genetic algorithms and the optimal allocation of trials. *SIAM Journal on Computing, 2*(2), 88–105.

Holland, J. H. (1975). *Adaptation in natural and artificial systems: An introductory analysis with applications to biology, control, and artificial intelligence.* Ann Arbor: University of Michigan Press.

Ingber, L., Petraglia, A., Petraglia, M. R., & Machado, M. A. S., et al. (2012). Adaptive simulated annealing. In *Stochastic global optimization and its applications with fuzzy adaptive simulated annealing* (pp. 33–62). Berlin/Heidelberg: Springer.

Kirkpatrick, S., Gelatt, C. D., & Vecchi, M. P., et al. (1983). Optimization by simulated annealing. *Science, 220*(4598), 671–680.

Oechssler, J., & Riedel, F. (2001). Evolutionary dynamics on infinite strategy spaces. *Journal of Economic Theory, 17*(1), 141–162.

Oechssler, J., & Riedel, F. (2002). On the dynamic foundation of evolutionary stability in continuous models. *Journal of Economic Theory, 107*(2), 223–252.

Ollivier, Y., Arnold, L., Auger, A., & Hansen, N. (2011). Information-geometric optimization algorithms: A unifying picture via invariance principles. arXiv preprint arXiv:1106.3708.

Rothlauf, F. (2006). *Representations for genetic and evolutionary algorithms*. Berlin/Heidelberg: Springer.

Schwefel, H.-P. (1993). *Evolution and optimum seeking: The sixth generation*. New York: John Wiley and Sons, Inc.

Wierstra, D., Schaul, T., Peters, J., & Schmidhuber, J. (2008). Natural evolution strategies. In *IEEE Congress on Evolutionary Computation, 2008. CEC 2008. (IEEE World Congress on Computational Intelligence)*, Hongkong (pp. 3381–3387). IEEE.

Chapter 3
Optimization

Abstract The optimization of a material requires tuning a single macroscopic property to take on an extreme value by manipulating microscale features. Here we show how artificial evolution can be used as an engine to accomplish this task, even in the absence of explicit relationships between micro-scale physics and bulk properties. As a particular example, we consider the task of designing the bulk mechanical properties of a granular aggregate by way of the constituent particle shape. We use artificial evolution to identify the particle shapes, that when aggregated, produce the stiffest and softest packings plus a packing that stiffens when compressed. These extreme materials are constructed and tested in the real world by 3D printing particles. Finally, we examine how optimization can be a means to extraordinary materials, the starting point for new scientific insights, or both.

3.1 Introduction

3.1.1 Optimizing Materials at the Micro-scale

With artificial evolution as an engine, we have the backbone for an automated materials design system that optimizes macroscopic properties. The next step becomes to identify which parameters the optimization engine should manipulate to achieve control over the bulk material response. At the micro-scale, differences in the shapes, interactions, and arrangement of constituent elements (molecules, atoms, grains, etc.) are the only ways to change a material's macroscopic qualities. Thus it stands to reason that a proof of principle optimizer should optimize materials using these microscopic properties as the mutation variables.

In our opinion, the most challenging microscopic variable to control is shape. Pairwise interactions can usually be specified with a handful of parameters, making even brute-force an option. Likewise, the material structure is a consequence of interaction types and preparation procedures. Thus, these factors can be parameterized or treated with interpolation functions, as seen in the literature (Torquato 2009; Jain et al. 2014). By comparison, shape is an inexhaustible parameter and it

M.Z. Miskin, *The Automated Design of Materials Far From Equilibrium*,
Springer Theses, DOI 10.1007/978-3-319-24621-5_3

is not clear how its role can be explored, let alone optimized. Thus, systematically optimizing with respect to shape is often considered infeasible, leaving the goal of a complete framework for automated materials design unfulfilled.

For several promising new materials platforms, this is particularly problematic. Increasingly, particle shape appears as a control knob for materials intrinsically out of equilibrium, like granular aggregates (Miskin and Jaeger 2013; Athanassiadis et al. 2014; Baker and Kudrolli 2010), as well as colloidal and nano-particle systems (Glotzer et al. 2005). Yet the standard approach is to fabricate first, and after the fact, identify functionalities. This tactic cannot be counted as rational design, and provides no clear path towards optimization. In short, the ability to optimize material behavior via constituent shape is not only an open problem, but could find use in materials systems spanning nine orders of magnitude in particle size.

We take on this challenge by considering not just how to optimize a material's properties, but how to do it by evolving particle shapes for a system far from equilibrium. Specifically, we consider the task of tuning the mechanical response of granular materials through particle geometry. As noted previously, granular materials are a particularly difficult system to design, because in contrast to nano-particles or colloids, there are no theoretical tools available to predict most basic properties (Jaeger et al. 1996; Duran 1999; Pena et al. 2007). Indeed, as shown in the next section, empirical relationships between particle shape and mechanical response in granular systems are rugged and complex, suggesting that even if a theory existed, it may prove too cumbersome to use for design.

3.1.2 *Shape in Granular Mechanics*

To get a sense of how shape impacts the macroscopic properties for a granular aggregate, compare the compressive stress-strain response for particles consisting of one, two and three equal-size rigidly bonded spheres (Fig. 3.1). In each case, the particles were randomly packed into a cylindrical enclosure and confined by equal pressure from all axes (Fig. 3.1a). This materials test, known as triaxial compression, is widely used for granular materials and standard in soil mechanics (Wood 1990).

The 2-sphere particles provide both a stiffer and significantly stronger packing than the single spheres, although the stress-strain curves otherwise appear similar (Fig. 3.1b). Some of this improvement is caused by the neck at the point where two spheres are joined, enabling some interlocking (Schreck et al. 2010; Galindo-Torres et al. 2009). However, by going to 3-sphere particles it becomes clear that the situation is more complex. For these particles, shape can be characterized by a single parameter, the opening angle at the central sphere, θ. Figure 3.1c shows simulation results for the average elastic modulus E as a function of θ. While the overall trend is intuitive, compact particle shapes giving rise to stiffer packings than extended rod-like shapes, it is remarkable how even modest changes in the opening angle produce significant changes in E. The resulting rugged landscape, full of local extrema, illustrates that, beyond corrugation originating from necks,

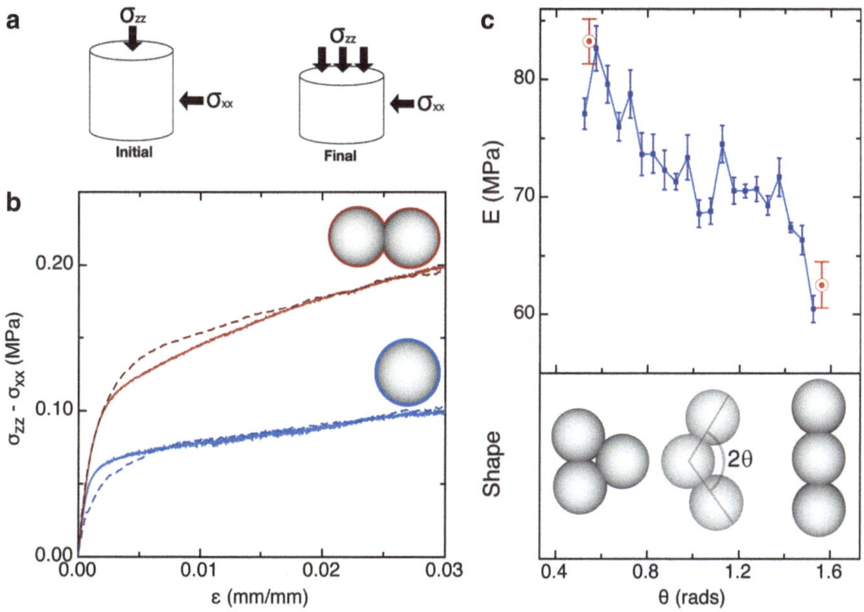

Fig. 3.1 The impact of particle shape on mechanical response (Miskin and Jaeger 2013). (**a**), Sketch of triaxial test procedure. A cylindrical volume of granular material is prepared by confining it with pressures $\sigma_{xx} = \sigma_{yy} = \sigma_{zz}$. This defines the initial condition of strain $\epsilon = 0$. The material is then compressed along the vertical, z direction. (**b**), Changing the particles from single spheres to dimers consisting of two rigidly connected spheres increases the stiffness and failure stress of the material significantly both in experiment (3D-printed particles, *solid curves*) and in simulations (*dashed curves*). (**c**), The rugged, non-linear relationship between shape and stiffness for particles made from three adjoined spheres. For such trimers the full range of shapes can be parameterized by the opening angle θ, as indicated. The *two red data points* correspond to the shapes shown below on the left and right. They have abscissa identified as extrema by our algorithm and experimentally obtained ordinates. Error bars correspond to the standard deviation from 10 (4) trials for the simulations (experiments)

a key role is played by the overall particle geometry. These numerical findings are confirmed by experiments on packings of 3D-printed copies of the granular molecules (Fig. 3.1b, c), confined and compressed under conditions similar to those of the simulations.

The intricate relationship between the geometry of individual particles and the properties of the aggregate emerging already for very simple shapes gives a hint of both the possibilities and the complexities. This can be contrasted with the relatively simple changes that result by changing material parameters. For example, if the friction stiffness is lowered from a physical value to zero, the stiffness of all the 3-ball particle packings decreases monotonically until they are within noise of one another (Fig. 3.2). In other words, when varying the effects of friction with a fixed shape, the general trend is a monotonic decrease in packing stiffness. Conversely,

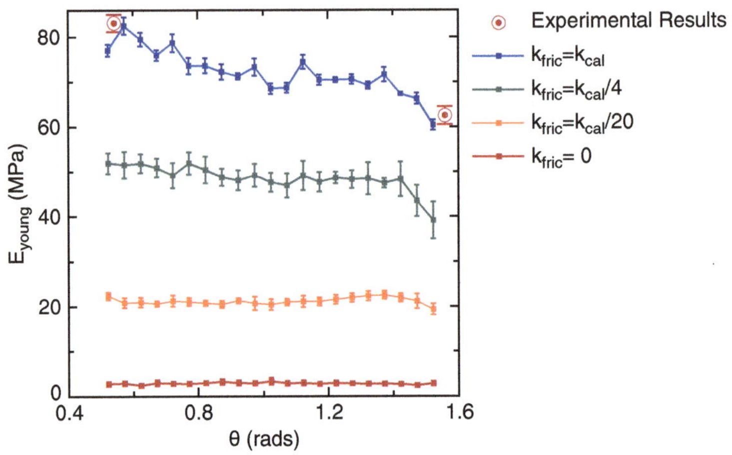

Fig. 3.2 The stiffness that results for granular packings made from three sphere shapes as the frictional stiffness is reduced relative to the normal stiffness of the particles (Miskin and Jaeger 2013). As the frictional force vanishes, the stiffness landscape transforms from a rugged, complex one with definable extrema to being essentially constant within the noise

varying shape at fixed frictional parameters produces a rugged and complicated landscape full of local extrema. Taken at face value, the hardest optimization problems surrounding granular materials then become tackling the immense, and vexing, search space of particle shape.

3.1.3 Optimizing Granular Materials via Shape

From a practical point of view, there are immediate benefits to optimizing aggregates since materials like sand, grain, and pharmaceuticals are essential in engineering and industry. In principle, the ability to rationally select granular particles to promote desired properties would find uses any time grains are synthesized directly for applications. For example, break waters and packing peanuts are both granular materials made from oddly shaped particles intended to dissipate bursts of energy. Clearly this represents a design goal, but in the absence of a far from equilibrium design method, the objects built to meet it were selected ad hoc.

The task of optimizing granular materials also connects with fundamental problems. In many ways, granular materials represent model systems for studying far from equilibrium physics (Jaeger et al. 1996; Liu and Nagel 2010). Thus, understanding the relationship between microscopic structure and macroscopic response could shed light on other amorphous systems like foams, dense colloids and molecular glasses. We believe generalizing results to other systems is easiest when studying configurations with extreme properties. By reasoning about shared features between extreme configurations, it becomes easier to separate the fundamental physics that produces large effects from the lower order components.

In the following sections, we work out a solution to the specific design problem of finding shapes that extremize the mechanical stiffness of a granular packing under compression. In the absence of a theory that unambiguously links particle shape to mechanical response, we turn to evolution and computer simulations to design extreme particles. As an integral part of this implementation, we introduce a representation for particle shape made of bonded spheres, or "granular molecules," that is easily mutatable as well as sufficiently general. Connecting these parts yields an automated design framework. We use this system to automatically discover the stiffest and softest shapes made from different numbers of spheres as well as an unconventional particle that becomes stiffer under compression. All of these results are verified by experiments on 3D printed versions of the designed shapes.

3.2 Experiment and Simulation Methodology

An essential component to our methodology lies in matching the fitness landscape from actual, physical experiments to the surrogate landscape generated by simulations. To achieve this, we established a methodology for generating reproducible triaxial test results, replicated it as closely as possible in the simulation procedure, and calibrated simulation parameters to measured physical parameters.

3.2.1 *Experimental Procedure*

We fixed all of our material parameters by 3D printing particles in UV cured hard resin (Objet Geometries, VeroWhite). Our particles have a characteristic size of 3 mm, and roughly 2,000 copies of each type of compound particle (100 mm^3 particle volume) were printed at a resolution of 50 μm. At this resolution, our printer (Objet Connex 350) guaranteed that, for all practical purposes, particles could be treated as identical objects.

To establish a reproducible initial state we adopted the following loading and pre-compression procedure for all of our triaxial tests. First, the sample is poured into a latex membrane 50 mm in diameter, 130 mm tall, 6 mm thick, capped on one end. The large aspect ratio is chosen here to decrease the relevance of friction with the end caps. Next, the sample is sealed on the top, and tapped. We lightly tap 50 times, invert the sample, and then tap 50 more times. The sample is then loaded into an Instron materials tester.

The tester holds both the top and bottom caps fixed in place, measuring the force needed to do so. We then pull a vacuum (typically 85 kPa) inside the sample to generate confining stress. The top plate is used to compress the sample until the stress on the top plate matches the confining stress. We check to make sure the measured stress is stable for at least 5 s, and then begin the compression test. If the stress is seen to fluctuate over this time, we decompress the sample slightly and then press back down until the stresses on all axes are matched again. During compression, samples are loaded at a rate of 10 mm/min to strains near 4 %. Since the deformation is relatively small, both stress and strain are calculated using the engineering conventions: we divide the measured force (displacement) by the initial sample area (length) to find the stress (strain) values.

3.2.2 Simulation Procedure

On the computer we simulate this procedure as closely as possible, using the PFC3D molecular dynamics software package from Itasca. Simulations are carried out on packings of roughly 2,000 particles compressed to 4 % strain. Isotropic random packings are generated via the radius expansion method. The packings are confined by cylindrical walls. To impose a specific stress state the walls of the cylinder are expanded and contracted, until the desired stress state is reached and stable. The compression test then begins at the same rate as in the experiments. During the test, the top and bottom plates are forced to reach specific values, whereas the cylindrical confining wall is allowed to expand or contract such that the packing is always maintained in a fixed confining pressure.

Particles can interact with one another by compression and friction. We model elastic compression as a linear spring with a fixed stiffness. We model friction as a linear spring force tangential to the normal with a spring constant different from the normal stress, unless the relative ratio of shear force to normal force exceeds a critical value (Cundall and Strack 1979). If this threshold is exceeded, the shear force is simply set in proportion to the normal force. The full details of the simulation model can be found in prior work (Potyondy and Cundall 2004). Altogether, the model has three free parameters, tunable to match experimental results: the two spring stiffness, and the Coulomb friction coefficient.

3.2.3 Calibrating the Simulation Parameters

In our experiments using 3D printed particles, we found that the force between two particles actually is approximated as linear in our operating regime, and that the effective spring constant scaled in proportion to the sphere radius. In other words, the normal force between the spheres is well approximated a linear spring. The value for the normal force spring constant was selected by 3D-printed spheres of different radii, compressing them until failure, and measuring force vs displacement (Fig 3.3).

Since the force scales linearly with compression and is proportion to the sphere radius, we believe that stresses originate from plastic yield onset. In this case the force is given by $f = \sigma_{yield} A$, where σ_{yield} is the yield stress and A is the contact area.

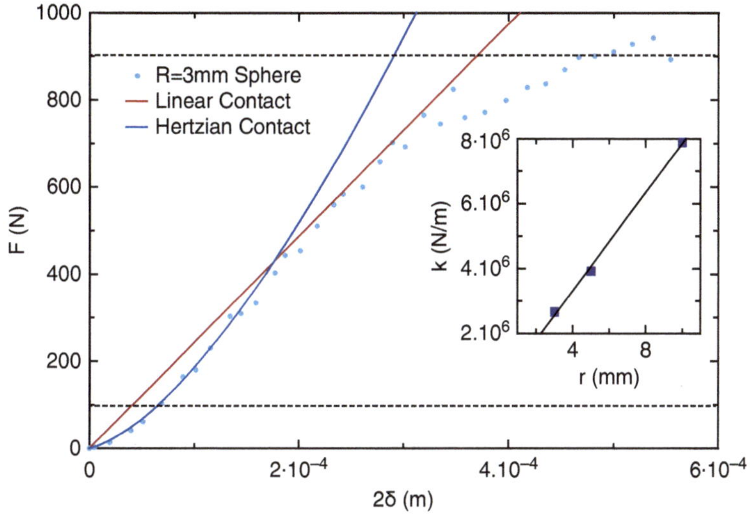

Fig. 3.3 Force displacement curve for a 3D-printed sphere (Miskin and Jaeger 2013). By compressing spheres of varying radii in our materials tester, we were able to find good materials settings for the normal force spring constant in our simulations (*red curve*). Note that the *dashed black lines* correspond to the largest and smallest force felt by a packing of trimers during a triaxial test at 85 kPa confining pressure. Thus, we anticipate that most of the force response is outside the Hertzian contact regime (*blue curve*). Also, note that the spring constant scales in proportion to the radius (*inset*), indicative of force arising from yield outside the linear elastic regime

For small indentations of a sphere, $A \sim 2\pi R\delta$ where R is the sphere radius and δ is the depth of indentation. Putting it together, the force law becomes $f = 2\pi\sigma_{yield}R\delta$. Note that the linear fit in the inset of Fig. 3.3 implies a compressive yield stress of 70 MPa, which is typical for hard plastic.

To pick a value for both the shear spring stiffness, and to select a failure coefficient, we compared simulated results to experiments for spheres and dimers. The parameters which gave the best agreement were then selected (Fig. 3.1).

We think this calibration procedure works well for two reasons. First, the order of magnitude for the shear spring constant has been set already by selecting the normal spring constant from experimental data. Because of this, the range of parameters for calibration has been dramatically reduced. Second, the shear stiffness spring value often plays little to no role in the final yield state. Instead the critical state is largely determined by the friction coefficient. Thus the parameters are effectively orthogonal, allowing the adjustment of one independent of the other. The final calibrated parameters are listed in Table 3.1

Finally, we note that the agreement between experiment and simulation will degrade if the strain is large. In this case, the change in area of the sample becomes significant and two sources of error arise. First the use of engineering stress becomes inappropriate, leading to a systematic overestimate in measured stress values. Second, since the simulations constrain the confining wall to remain

Table 3.1 Calibrated parameters used for simulations of a triaxial compression test

Compressive modulus	70 MPa
Density	1200 kg/m^3
Normal to shear stiffness	5
Friction coefficient	0.25

cylindrical, inhomogeneous deformations of the boundary will result in discrepancy. We ignore these effects in our work since we operate in the low strain regime; however future studies, in particular those interested in optimizing failure stress, will have to account for these problems explicitly.

3.3 Developing a Representation for Shape

3.3.1 The Basic Blueprint Rules

To successfully optimize particle shape, we need to create a representation for particles that is both unrestrictive and amenable to mutation. To meet this need, we developed a shape representation that we call a blueprint. In the same way a picture is represented by pixels, we represent an arbitrary shape as a volume composed from smaller building blocks. Abstractly, we define a blueprint as a list of bearings written in sequential order specifying where to place each building block as the compound object is being constructed. For this list of directions to define a single object, we add the rule that each new part is always added along its bearing at the furthest distance that leaves it still connected to the structure built during the previous steps.

To demonstrate the principle, in the following we use equal-size spheres as the building blocks, bonding them together to form rigid compound particles we term granular molecules. This use of glued, slightly overlapping spheres extends earlier work to represent non-spherical particles in simulations of granular friction (Galindo-Torres et al. 2009; Poschel and Brilliantov 2003). It has the computational advantage that contacts among neighboring molecules in the packing involve only a single type of sphere-sphere interaction (Kodam et al. 2009).

We can use each blueprint as a genotypic expression for shape, and evolve populations by mutating the bearing angles. Each genotype can then be mapped uniquely to a particular phenotype for testing, as illustrated by Fig. 3.4. Importantly, this approach also minimizes the dimensionality required by the parameter space: since each block is constrained to remain connected to the built object, only two variables rather than three spatial coordinates are needed to uniquely specify its location.

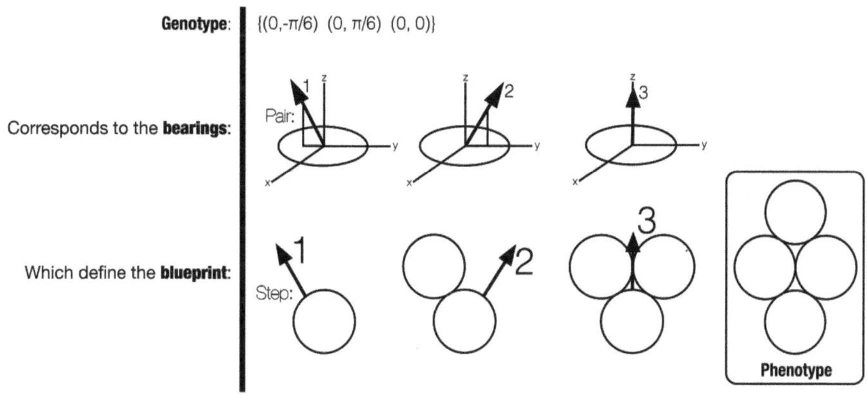

Fig. 3.4 Representing particle shapes with blueprints (Miskin and Jaeger 2013). To construct a single shape from a genotype of angle pairs (n = 4 particle represented by three angle pairs in the example shown here), the angles are first mapped to bearings in spherical coordinates. A first sphere is then placed. This sphere will act as the origin and all bearing vectors will be drawn out originating from the center of this sphere. The remaining spheres are placed iteratively as follows. For each bearing, a ray is drawn that emanates from the center of the origin sphere and is parallel to the bearing. If a sphere is then slid through its center along this ray from infinitely far away towards the origin, then, the location it first comes in contact with the built structure defines the furthest distance along the bearing that still leaves it connected to the previously built shape. By taking each bearing in sequence, and placing a sphere at these distances, a phenotype shape can be constructed and used in simulations

3.3.2 Benchmarking the CMA-ES Using Blueprint Rules

In benchmark tests, we found this shape representation highly efficient. For instance, using the CMA-ES as an optimizer we were able to evolve initial, random sphere configurations into rods (Fig. 3.5a), rings (Fig. 3.5b), deltahedra, or cubes in typically no more than a few hundred generations.

In particular, the task of minimizing the sum of the distances to the center of mass became a good benchmark to study how the algorithm scales in a complex shape space. This problem was previously studied by multiple authors, and optimal configurations are well known (Arkus et al. 2009; Sloane et al. 1995). For objects made with $n < 13$ spheres, the correct solutions are deltahedra, i.e. polyhedra with triangular faces. Finding these optima is remarkably challenging since the number of local minima increases exponentially with the number of elements (Arkus et al. 2009). However, using the CMA-ES, we were always able to find the exact solutions when searching in the regime, $n = 3$–5 (Fig. 3.6c) with relative errors at 10^{-14} (Fig. 3.6b). Furthermore, a solution within a percent or less of the global optimum is usually obtained within the first 100 generations.

When the number of spheres in the search space exceeds 7, the probability of finding the global optimum moves from 1.0 to 0.5 (Fig. 3.6c). In practice, this is a small hurdle. First, we note that the solutions obtained are still excellent (relative

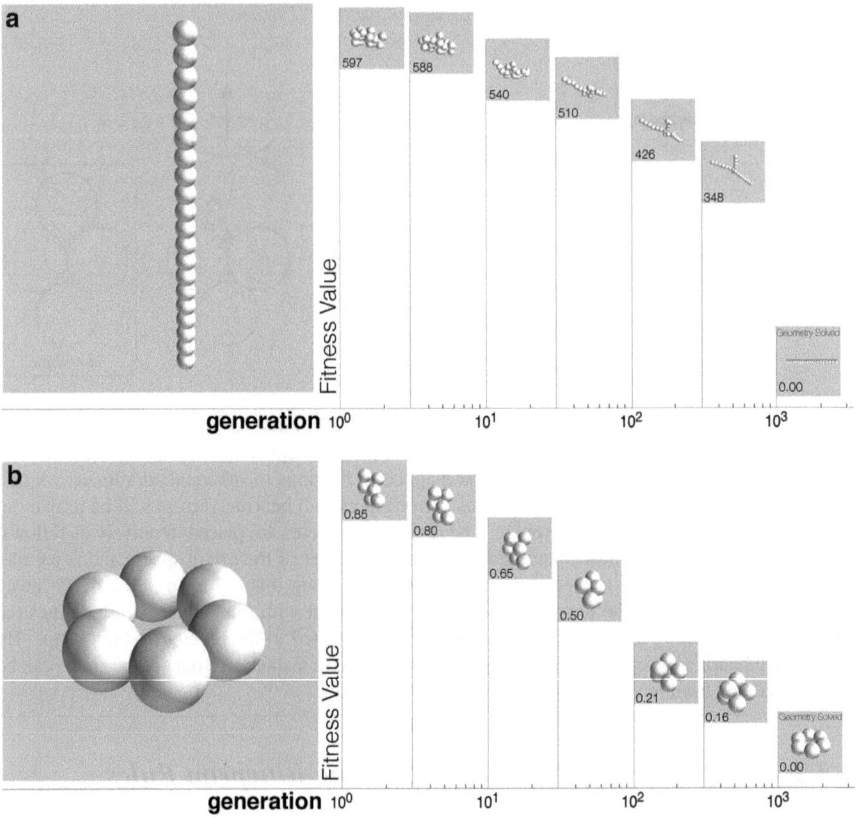

Fig. 3.5 Evolving rods and rings (Miskin and Jaeger 2013). Our algorithm can evolve shapes in spaces of large dimensionality quite efficiently. Here we used it to evolve a 20-sphere rod, corresponding to a 2x(20-1) = 38 dimensional search space, with convergence in only 1,000 iterations (**a**). Additionally, the algorithm can explore the impact of both topology and geometry as illustrated by its mutation of randomly distributed spheres into a closed ring (**a**). The fitness functions used to generate the figures were "maximize the sum of the distances to the center of mass (**a**)" and "minimize the sum of the standard deviations to the center of mass plus the distances from the plane z = 0 (**a**)"

errors $\sim 10^{-2}$). Second, in this case, the decline does not appear as a trend for larger dimensions, but rather saturates around 0.5. That said the effect can be almost completely mitigated by simply running the optimization from a handful of fresh starting configurations. We also note that this small decline is problem specific. For instance, if we change the fitness metric to maximize the sum of the distances to the center of mass, our algorithm always identifies the correct solution of rod geometries, even when the number of building blocks is as high as 20 spheres (dimension 40) (Fig. 3.5a).

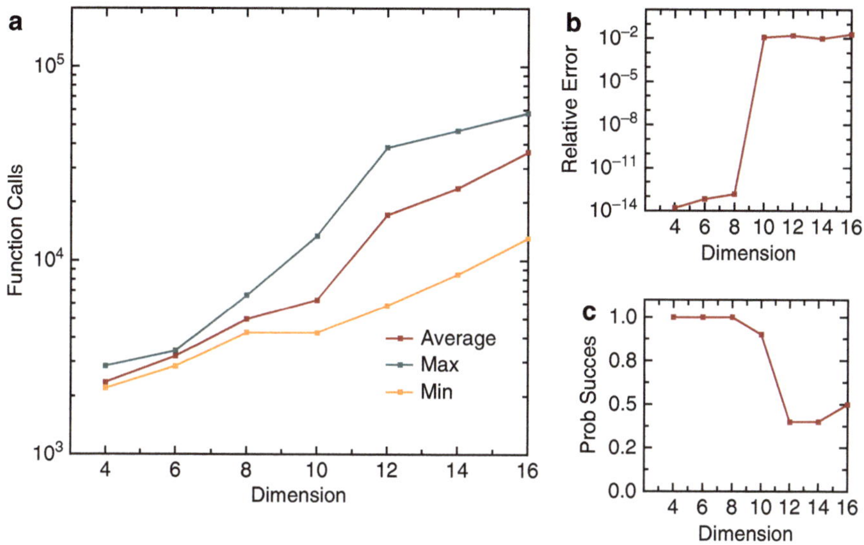

Fig. 3.6 Minimizing the sum of the distances to the center of mass for clusters of spheres via the CMA-ES (Miskin and Jaeger 2013). The hardest problem we posed to the CMA-ES was finding the shapes that minimize the sum of the distances to the center of mass. For this problem, the number of local minima grows exponentially with the problem dimension, making it particularly rugged. We allow the optimizer to run until the variation in fitness is less than 10^{-15} and plot the number of function calls (**a**) needed to identify a compact shape and the solution quality (**b**,**c**) against the dimensionality of the search space. We note that for dimensions less than 10, the CMA-ES always finds the global optimum within a few thousand function calls. Above dimension 10, the probability of success decreases, but not systemically. Instead it simply reduces from 1 to about 0.5. Thus rerunning the optimization only a few times is sufficient

3.3.3 *The Generalized Blueprint Rules*

The blueprint approach can be easily extended to different building block types. Indeed, since objects are fixed together at points of contact along the bearing, the same rules can be used irrespective of whether or not the building blocks are spheres. For instance, using cubes might prove useful in stochastic assembly (Tolley and Lipson 2011). Spherical building blocks can also be used directly for energy or moment optimization of colloidal particle configurations (Arkus et al. 2009; Hoy et al. 2012). Alternatively, keeping spheres as the building blocks, the lines adjoining overlapping spheres could be used to define beams, and the same blueprint could be mapped to truss structures.

A more advanced extension would be to use sub-assembled objects as building blocks thereby defining a hierarchy of blueprints. For example, the basic blueprints defined in the prior section could be used to build up a small collection of shapes. These shapes can then be used as the building blocks for a higher level blueprint constructor. The procedure is sketched in Fig. 3.7.

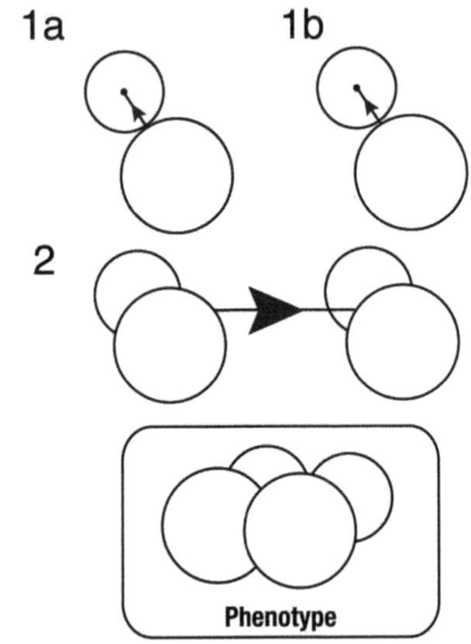

Fig. 3.7 Blueprint rules do not specify the shapes of the objects they assemble. Thus they can even assemble objects built up using blueprint rules. In this example, shapes 1a and 1b are assembled using the pictured 1-bearing blueprints. They are then taken as input and assembled at stage 2 by adjoining them along a new blueprint, producing a final object of 4 bonded spheres

This hierarchy facilities dividing complex objects into successively smaller, more manageable parts, much like a grammar does for sentences. In this sense, we can think of blueprint rules as a "language for shapes:" angles form the basic elements like words, and hierarchies in the assembly process act like punctuation.

In this work, we do not peruse this possibility. However we think it is worth describing both for its intellectual merits and because, in future work, it seems like the most natural route for extending blueprint rules to larger, more complex objects. We can envision a scenario in which individual optimizers act at different levels of the hierarchal build space. Each individual optimizer then sees a simpler, more tractable task of learning to optimize its small component conditioned on the angles above it in the hierarchy. Adding this structure would in principle reduce the problem complexity without reducing the space of build-able objects.

3.4 Optimization Results

At this stage, we have the two essential ingredients to optimize granular materials with artificial evolution: we can evaluate the quality of objects compared to a design goal using the calibrated simulations and we can parameterize shape as a set of mutable angles with the blueprint rules. It is now a straightforward exercise to connect the two through an optimization algorithm and begin evolving.

3.4.1 The Stiffest and the Softest Packing Shapes

Applying an evolutionary approach to jammed particle packings, we initially consider two related goals: to find the particle shapes that produce the stiffest response and the softest response to compressive loading. That is, shapes which produce packings with the largest (smallest) elastic modulus near zero strain.

For granular molecules consisting of three spheres, the search converged on the configurations producing the extrema in E found earlier by varying θ directly (Fig. 3.1c), in excellent agreement also with the experimental data (shown in red). For 4-sphere molecules, Fig. 3.8a shows the history of the parent shapes leading to the convergent results for each goal.

After the CMA-ES converged on the optimized $n = 4$ shapes, we used 3D printing to construct physical packings and verify the results with triaxial tests. In Fig. 3.8b we plot the resulting stress-strain curves. The general agreement of the simulations with the experiments asserts that rejected shapes were, indeed, suboptimal. Further, the significant differences between the shapes producing the stiffest and softest response suggest that the evolutionary algorithm makes full use of the available configuration space. Figure 3.8c emphasizes this point with a histogram of simulated elastic moduli for $n = 4$ genotypes using random blueprints. The stiffest and softest moduli are found 4 and 6 standard deviations from the mean, respectively, demonstrating that the evolutionary approach to granular design is both accurate and efficient.

Repeating this process to uncover the stiffest and softest shapes up to $n = 5$ causes clear motifs underpinning the optimized designs to emerge (Fig. 3.9). Over this range in n, the softest shapes discovered are always linear molecules. By contrast, the stiffest shapes are relatively compact. Specifically, we find the n stiffest shape can be constructed by taking the most compact $n + 1$ shape, i.e., the shape whose configuration of $n + 1$ spheres minimizes the sum of their distances to the center of mass, and removing a sphere (Fig. 3.10). For example, the most compact shape for $n = 5$ is the triangular bipyramid. Removing one sphere produces the slightly bent, approximately rhombic shape, discovered independently as the $n = 4$ stiffest shape. We believe that the lucidity of these motifs makes them useful for applications and provides a first step toward the rational selection of granular shapes.

We emphasize that particle shapes producing the stiffest load response are not necessarily the ones that pack most densely. Granular materials resist stress because particles are impeded from sliding past one another, and thus the nature of the local contacts matters as well. The socket-like hole produced by removing one sphere from the most compact $(n+1)$-shape appears to provide a local packing environment where two contacting n-shapes can share a sphere and lock together. For relative movement, the lock must first be undone. Linear granular molecules, on the other hand, can only form weaker connections and contacts between overlapping rods are more susceptible to torque.

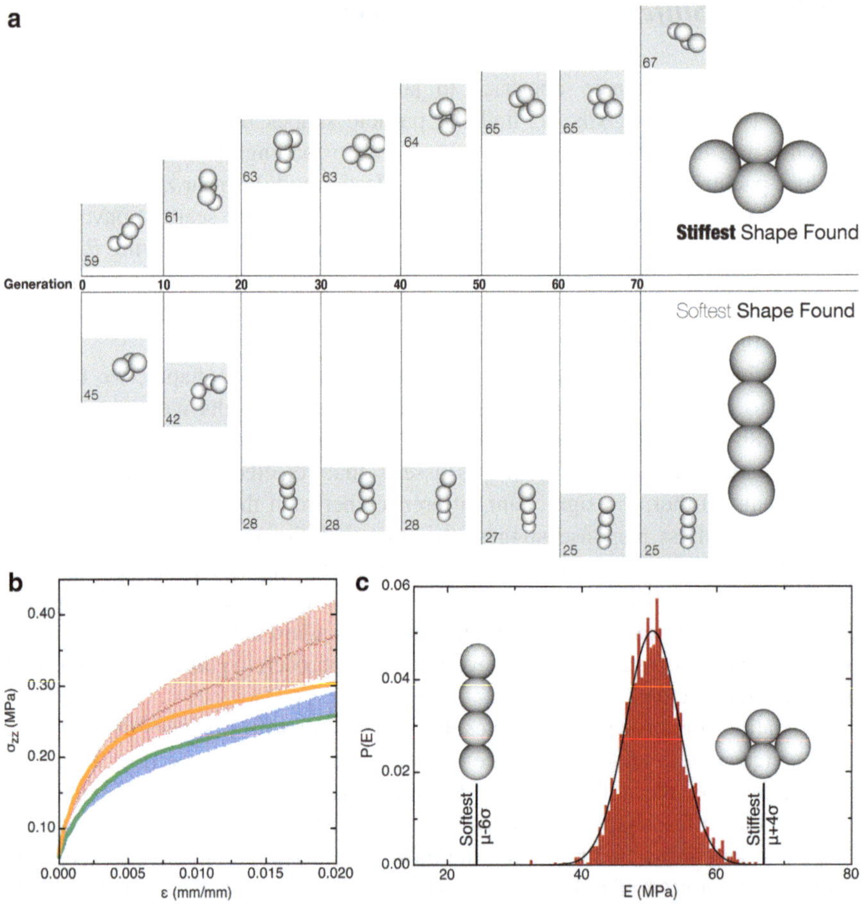

Fig. 3.8 Evolving particle shapes to obtain the stiffest and softest packings (Miskin and Jaeger 2013). Shown are examples using particles built from $n = 4$ spheres. Starting with a random shape, the evolution process is iterated over 70 generations, converging to optimized shapes. (**a**), Snapshots of those parent shapes that produce, in the generations indicated, the packings with the largest (smallest) elastic modulus. The vertical position of the images is scaled to the value of E, which is also indicated in the lower left corner of each gray box. (**b**), Comparison between packings of 3D-printed particles and predicted stress-strain curves shows excellent agreement in the low-strain regime where the elastic modulus was optimized. At larger strains the data for the stiffest (*orange*) and softest (*green*) shapes start to deviate from the experimental results (*red* and *blue*, respectively; error bars correspond to the standard deviation from four independent experiments). (**c**), Histogram of elastic moduli obtained with random blueprints. The line is a fit to a Gaussian with mean 50 MPa and standard deviation 4 MPa. Comparison with the shapes found in (**a**) demonstrates efficiency of the evolutionary approach over random search: to discover the stiffest and softest shape would take roughly 15,000 and 500,000,000 guesses, respectively, rather than the 600 actually used

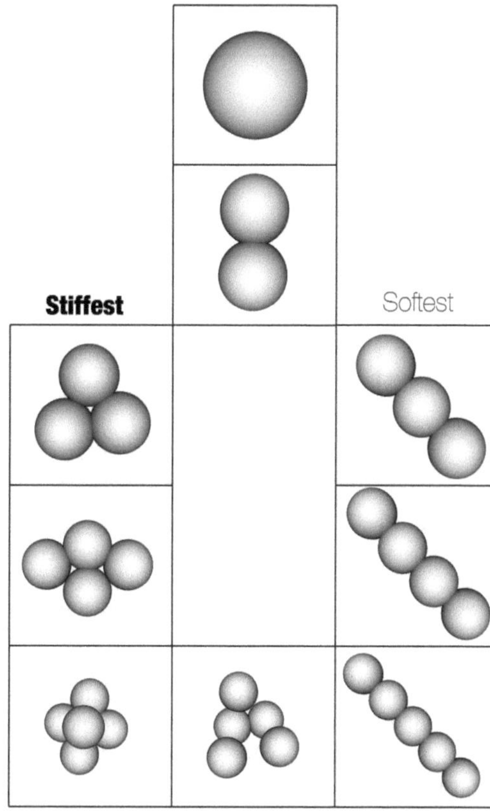

Fig. 3.9 Stiffest and softest particle shapes built from spheres up to n = 5 (Miskin and Jaeger 2013). Two clear motifs emerge. First the softest shapes discovered are always rods. Second, the stiffest n-shape can be constructed by taking the most compact (n + 1)-shape and removing a single sphere. Note that the stiffest shape for n = 4, obtained from the n = 5 triangular bipyramid, is not as planar as it appears in the top view shown, but slightly bent and 'boat-shaped'. In the center we also include our n = 5 strain-stiffening particle; note the asymmetry in its sphere arrangement

3.4.2 Discovering Strain-Stiffeners

The link between contact locking and stress response invites more exotic design goals. For instance, suppose a material was initially in a state characterized by weak particle contacts, but as these contacts were broken, new contacts formed that were stronger. Such material would stiffen the further it was compressed, in contrast to the strain softening common for granular materials (see Figs. 3.1b and 3.8b). We used our algorithm to search for appropriate shapes, i.e. shapes producing the largest second derivative of stress with respect to strain. The result was a particle that, with just five constituent spheres, inverts the usual stress response, producing a concave-upward dependence for $\sigma(\epsilon)$ (Fig. 3.11a). This wishbone-like shape produces packings that exhibit a low modulus, yet do not fail up to 10 % strain and sustain stresses significantly above the confining pressure (Fig. 3.11b). The evolutionary strategy in this case led to the discovery of a unique, asymmetric particle that achieves not only a very different load response but in fact a qualitative change in the stress-strain relationship.

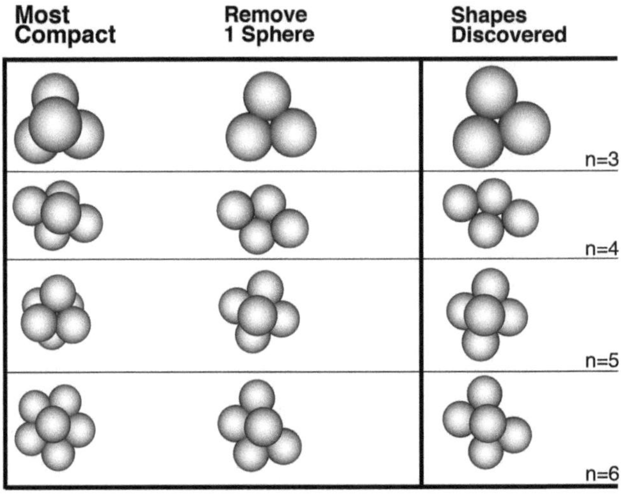

Fig. 3.10 A motif among the stiffest particle shapes (Miskin and Jaeger 2013). Removing one sphere from the most compact shape for n + 1 spheres (*left column*) produces shapes of n spheres (*center*) almost identical to those located independently by our algorithm (*right column*). It is, however, unknown which sphere should be removed to generate the stiffest packing

Strain-stiffening was recently found also in packings of granular polymers (linear chains of flexibly linked beads) if the polymers were long enough (Brown et al. 2012). Here, however, the wishbone-like granular molecule in Fig. 3.11a is not only rigid but also smaller, i.e., consists of fewer spheres, than granular polymers that strain stiffen. This relatively small molecule achieves remarkable self-confinement under load: it can sustain stress levels 10 times larger than the confining pressure (Fig. 3.11b). We also note that these packings exhibit a low modulus, yet do not fail up to 10 % strain, an order of magnitude larger than simple spheres under the same conditions.

3.5 Summary

Optimizing granular materials with a computer algorithm meets the key requirements of a machine that designs materials far from equilibrium. By machine we mean that once a user defines the simulations and the design goal, an algorithm takes over to automatically make the discoveries. First, this machine was used to find the families of stiffest and softest packing shapes. Then, it found a particle that stiffens under compression. And once this machine was built, all of these tasks were accomplished without any more user input than an objective.

Some of the results it produced are easier to rationalize than others. Chiefly, the strain-stiffening particle is harder to interpret than the stiffest and softest packing

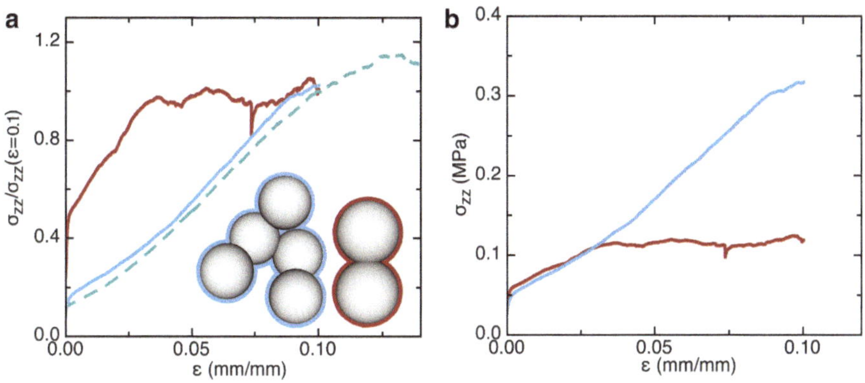

Fig. 3.11 Discovering a strain-stiffening particle (Miskin and Jaeger 2013). (**a**), Stress-strain curves for n = 5 particle shape designed to strain-stiffen (simulation: *green-dashed*; experiment: *blue-solid*). The confining pressure $\sigma_{xx} = \sigma_{yy}$ for these experiments had to be reduced to 35 kPa to prevent damage to the wishbone-shaped, 3D-printed particles. Therefore, we normalized the stress axis to compare simulations (calibrated for $\sigma_{xx} = 85$ kPa) to experiments. Note the presence of a new inflection point, which leads to positive curvature, in contrast to all other shapes that strain-soften, exemplified here by experimental data on dimer particles under 35 kPa confinement (*red*). (**b**), Packings formed by the strain-stiffening shape are strong as well as tough. As the non-normalized data show, the failure stress is roughly a factor 3 larger than for dimers under the same conditions, and it exceeds the confining pressure by an order of magnitude. This contrasts with packs of individual spheres, where the failure stress is comparable in magnitude to the confining pressure (Fig. 3.1b). At the same time, the strain-stiffening packings withstand much larger strains before failing (10 % compared to no more than a few percent for packings of dimers and even less for single spheres (see Fig. 3.1b)). These results demonstrate that our approach can control not just first, but also second order stress-strain effects

shapes. However, even features in the simplest resulting shapes are arguably too subtle to guess a priori. This is one of the strengths of automated design. Unlike serendipitous discovery, it is not dependent upon nor biased by prejudices or intuitions. Instead, it operates purely on an empirical level. Thus the final results are not directly required to be easy to conceptualize: they are only required to be good.

As a consequence of mechanizing this process, the results are left to the user to interpret. To the extent that a user cares only about applications, having an optimized shape may be completely satisfactory. Indeed the shapes in Fig. 3.9 could be used in and of themselves as optimal materials. Beyond that, automated design can afford even more significant flexibility and control. When $\sigma(\epsilon)$ is expanded as a Taylor series in strain, our results demonstrate that shape, for given boundary conditions, can tune both the first (Fig. 3.8) and second order (Fig. 3.11) terms. Past evolving shape with respect to only certain features of $\sigma(\epsilon)$, such as E, a straightforward extension could find the best match to a full curve.

Beside practical uses, we might find the results interesting for fundamental reasons. Having optimized objects for reference is like having hints at what matters: it makes guessing a hypothesis that generalizes the results far easier than relying on imagination alone. In addition to a tool for engineers, the shapes in Fig. 3.8 also form the basis for an inductive hypothesis about stiffness in granular materials. Alternatively, the strain-stiffening particle, by virtue of its unusual behavior, poses immediate questions by itself about mechanics in the granular state. To the extent that the user wants to explain the solutions, design by optimization can become the starting point for new science.

Finally, the kinds of systems explored by this machine are left at the discretion of the user. In many ways, the success of the optimizer is independent of whether or not it is solving a problem defined by simple geometric features, as it was during benchmarks, or solving a granular simulation, or simulations of colloids or nano-particles. While physically and computationally, these problems are all very different, they do not require any careful adaptation of the approach here to be attempted. In fact, switching between running benchmarks to optimizing via simulations only required us to change one line of code.

Along these lines, future work may consider adding degrees of freedom by treating the number of building blocks or their size as directly mutable parameters, or by co-evolving within the same packing environment particle shapes that emphasize particular behaviors, like those in Fig. 3.9. As discussed so far, only steric interactions between granular molecules were considered, but additional magnetic or electrostatic interactions could be explored by incorporating site-specific 'functional groups' into the molecules. And if each of these avenues is explored, we believe that the limits to optimized design will be set only by the limits to computation.

References

Arkus, N., Manoharan, V. N., & Brenner, M. P. (2009). Minimal energy clusters of hard spheres with short range attractions. *Physical Review Letters, 103*(11), 118303.

Athanassiadis, A. G., Miskin, M. Z., Kaplan, P., Rodenberg, N., Lee, S. H., Merritt, J., Brown, E., Amend, J., Lipson, H., & Jaeger, H. M. (2014). Particle shape effects on the stress response of granular packings. *Soft Matter, 10*(1), 48–59.

Baker, J., & Kudrolli, A. (2010). Maximum and minimum stable random packings of platonic solids. *Physical Review E, 82*(6), 061304.

Brown, E., Nasto, A., Athanassiadis, A. G., & Jaeger, H. M. (2012). Strain-stiffening in random packings of entangled granular chains. *Physical Review Letters, 108*, 108302.

Cundall, P. A., & Strack, O. D. (1979). A discrete numerical model for granular assemblies. *Geotechnique, 29*(1), 47–65.

Duran, J. (1999). *Sands, powders, and grains: An introduction to the physics of granular materials*. Berlin: Springer.

Galindo-Torres, S. A., Alonso-Marroquin, F., Wang, Y. C., Pedroso, D., & Munoz Castano, J. D. (2009). Molecular dynamics simulation of complex particles in three dimensions and the study of friction due to nonconvexity. *Physical Review E, 79*, 060301.

Glotzer, S. C., Horsch, M. A., Iacovella, C. R., Zhang, Z., Chan, E. R., & Zhang, X. (2005). Self-assembly of anisotropic tethered nanoparticle shape amphiphiles. *Current Opinion in Colloid and Interface Science, 10*(5), 287–295.

Hoy, R. S., Harwayne-Gidansky, J., & O'Hern, C. S. (2012). Structure of finite sphere packings via exact enumeration: Implications for colloidal crystal nucleation. *Physical Review E, 85*(5), 051403.

Jaeger, H. M., Nagel, S. R., & Behringer, R. P. (1996). Granular solids, liquids, and gases. *Reviews of Modern Physics, 68*(4), 1259–1273.

Jain, A., Bollinger, J. A., & Truskett, T. M. (2014). Inverse methods for material design. *American Institute of Chemical Engineers Journal, 60*(8), 2732–2740.

Kodam, M., Bharadwaj, R., Curtis, J., Hancock, B., & Wassgren, C. (2009). Force model considerations for glued-sphere discrete element method simulations. *Chemical Engineering Science, 64*(15), 3466–3475.

Liu, A. J., & Nagel, S. R. (2010). The jamming transition and the marginally jammed solid. *Annual Review of Condensed Matter Physics, 1*, 347–369. Annual Reviews.

Miskin, M. Z., & Jaeger, H. M. (2013). Adapting granular materials through artificial evolution. *Nature Materials, 12*(4), 326–331.

Pena, A. A., Garcia-Rojo, R., & Herrmann, H. J. (2007). Influence of particle shape on sheared dense granular media. *Granular Matter, 9*(3–4), 279–291.

Poschel, T., & Brilliantov, N. V. (2003). *Granular gas dynamics* (Vol. 624). Berlin/Heidelberg: Springer.

Potyondy, D. O., & Cundall, P. A. (2004). A bonded-particle model for rock. *International Journal of Rock Mechanics and Mining Sciences, 41*(8), 1329–1364.

Schreck, C. F., Xu, N., & O'Hern, C. S. (2010). A comparison of jamming behavior in systems composed of dimer- and ellipse-shaped particles. *Soft Matter, 6*(13), 2960–2969.

Sloane, N. J. A., Hardin, R. H., Duff, T. D. S., & Conway, J. H. (1995). Minimal-energy clusters of hard-spheres. *Discrete and Computational Geometry, 14*(3), 237–259.

Tolley, M. T., & Lipson, H. (2011). On-line assembly planning for stochastically reconfigurable systems. *International Journal of Robotics Research, 30*(13), 1566–1584.

Torquato, S. (2009). Inverse optimization techniques for targeted self-assembly. *Soft Matter, 5*(6), 1157–1173.

Wood, D. M. (1990). *Soil behavior and critical state soil mechanics*. Cambridge: Cambridge University Press.

Chapter 4
Inverse Problems

Abstract Suppose that for a particular material, a given microscopic property is known to impact a given bulk response. How can one construct simple rules that invert this relationship to identify the microscopic configurations that yield specific, desired responses? This is what we term an inverse problem: find a simple procedure that can be used by itself to connect at least one microscopic configuration to any possible bulk behavior. In this chapter we show how to solve inverse problems by way of design rules. As a specific example, we consider how particle shape impacts the role of packing density in a granular aggregate. We produce design rules that can be used to link one particle shape to any given packing density that falls within an achievable range and the results are verified by experiments on 3D printed particles. Finally we discuss how the results achieved here are a prototype for solving broad classes of design problems related to a variety of complex materials.

4.1 Introduction

4.1.1 Defining Inverse Problems

Optimization is a singular design process: it takes the range of possible material responses that can result from a parameter, and compresses it to a single input-output pair that is labeled the best. Consequently, the results of an individual optimization can be hard to generalize, leaving open the broader issue of finding materials with arbitrary properties. Does one have to optimize again, searching for the material that best meets each individual specificity?

This general scenario is what we would call an inverse problem. Finding a solution means finding a middle ground of information between the complete, incomprehensible search space and the solitary results of optimization. What is needed here is a moderate amount of compression: just enough information to rationally link configurations to responses, but not so much as to make the process complicated.

As a definite example, consider packing density in a granular material. Different particle shapes, when poured into a container, will pack randomly to different densities (Baker and Kudrolli 2010; Donev et al. 2004; Wouterse et al. 2007; Jaoshvili et al. 2010). One can even use the optimization method from the prior

M.Z. Miskin, *The Automated Design of Materials Far From Equilibrium*,
Springer Theses, DOI 10.1007/978-3-319-24621-5_4

chapter to find the shapes that pack the most/least densely. Yet for an inverse problem one would have to answer questions like "which particle shapes pack randomly to a density of X," where X is any number that falls between the minimum and maximum densities. Technically, it is possible to optimize repeatedly and try to find a shape for every given X, but this is computationally burdensome, and hopefully unnecessary.

4.1.2 Design Rules as a Solution

For a given inverse problem, we believe a true solution should come in the form of 'design rules.' By this, we mean a set of instructions that can be used in and of themselves to easily associate microscopic features to any macroscopic behavior that falls within an achievable range. We contrast this with seeking a single object that solves a single inverse problem. We seek rules that can be followed to generate solutions to all the inverse problems that can be posed within a specific class.

To make these definitions clearer, return to the example of finding particle shapes that pack randomly to a density of X. Here the microscopic feature to control is particle shape while the macroscopic response to design is packing density. Suppose, as in the prior chapter, particle shapes are represented as bonded spheres, but the sphere radii are now allowed to take on arbitrary values. Solving the inverse problem then requires finding a collection of shapes that includes at least one that will fill to each possible density. Furthermore, finding a design rule is equivalent to finding a set of instructions that can be used to easily and rationally generate the shapes in this collection.

One kind of design rule can be built directly through the shape representation. Bonded spheres, generalized to variable sphere radii, are a design space in which any one shape can be morphed into another. By extension, any response can be morphed from one to another, provided the connection between shape and response is continuous. Thus for particle packing, a path of shapes that leads through response space from the minimum response to the maximum response can function as an inverse problem solution. This path effectively operates as a 'design rule solution,' at least to the extent that the range spans the min and max response, and that the path is easy to interpret.

This process is illustrated schematically in Fig. 4.1. Suppose, for example, that a cubic granular molecule needed to be morphed into a sphere. While there are infinitely many ways to accomplish this goal, a direct and convenient approach is to send the radius of each sphere in the cube to zero, one at a time in a specified order, until a single sphere remains. In other words, one labels each of the spheres in the cube arbitrarily, uses the labels to define an order, and then shrinks the spheres one at a time until a single sphere remains. The advantage of this procedure is that once the order of elimination is fixed, a single parameter, i.e. the sum of all sphere radii in the molecule, specifies each intermediate shape uniquely. By making a plot comparing this 'shape index' to an aggregate property, designated in Fig. 4.1 by Γ,

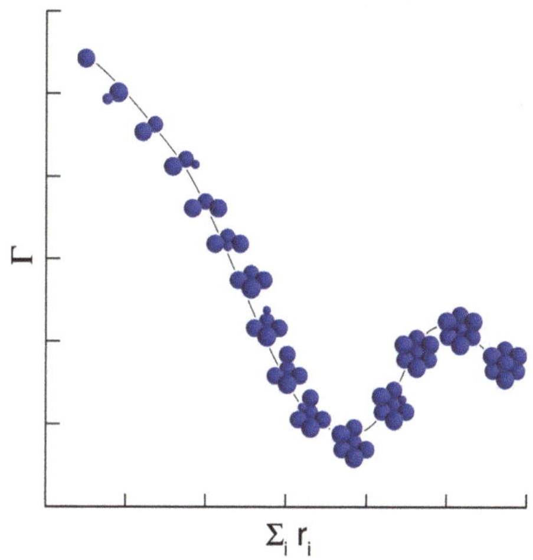

Fig. 4.1 A schematic illustrating how a series of deformations in shape can lead to a path of different aggregate responses Γ (Miskin and Jaeger 2014). In this case, a cube-shaped granular molecule is being transformed into a sphere by systematically sending the constituent sphere radii, r_i, to zero, until a single sphere remains. Given any specified order for removing the spheres, each shape in the path is uniquely identified by the sum of all radii in the granular molecule. Thus, if desired, this graph could be used to solve an inverse problem for the aggregate property Γ: simply draw a horizontal line from the y-axis to the intercept with the response space graph to locate the appropriate shape

a graphical map emerges that links any response in its range to at least one shape. Because the shape indices effectively encode the particle geometries, this map is a design rule that functions as an independent inverse problem solution.

With this in mind, the best design rule is the one that extends the range of responses to be as large as possible. This rephrases the inverse problem as an optimization problem that can be solved via artificial evolution: find the shapes that produce the minimum and maximum responses. The design rule then emerges from considering the path, as in Fig. 4.1, that continuously transforms these two extreme shapes into one another.

In the next section we work through the specific example of finding particle shapes that pack randomly to a density of X. Since packing fraction is known to depend on both shape and on the role of friction, we consider both packings that are poured in place and packings that are tapped until reaching an asymptotic density. In simulations, we approximate these loading protocols by pouring particles into a box under gravity with frictional and frictionless contact force models. We show not only how to find the minimum and maximum packing density shapes, but then how to continuously transform one into the other. By simulating these deformed shapes, we build up a design rule and use it to link a particle to a randomly selected packing density.

4.2 Experiment and Simulation Methodology

4.2.1 Experimental Procedure

Because our design problem again focuses on particle geometry, it is essential to perform experiments in which shape is the sole variable. This means that material parameters, loading procedures and boundary conditions must be standardized to ensure agreement with simulations.

Particle configurations are again 3D printed in UV cured hard resin (Objet Geometries, VeroWhite) at a size of 3 mm. Roughly 5,000 copies of each type of compound particle were printed in high resolution (50 μm) to make packings.

Next we standardized a definition for volume fraction. We wanted to reduce variability from boundary effects as much as possible, and to do so, we poured samples into different sized and shaped beakers and extrapolated the results to an infinite system size. Our extrapolation method was based on prior tomography results that revealed walls produce a boundary layer of fixed length, l, in which the packing fraction deviates substantially from that of an infinite system (Suzuki et al. 2008).

For cylindrical containers, this fixes the dependence of measured packing fraction against system size to functions of the form $\phi_m = \phi_b(1-(1-\frac{a}{R}l)^2)+\phi_c(1-\frac{a}{R}l)^2$ where ϕ_b is the packing fraction in the boundary layer, ϕ_c is the packing fraction in the core region, a/R is the ratio of the largest sphere radius in a granular molecule to the container radius, and l is the boundary layer size, measured in units of a. For spherical containers, the formula is essentially the same but with a different exponent: $\phi_m = \phi_b(1 - (1 - \frac{a}{R}l)^3) + \phi_c(1 - \frac{a}{R}l)^3$

We use this function to fit measured packing fractions against system size. To remove dependence on boundary geometry, we performed the analysis by using different shaped containers. Specifically, we made measurements using both spherical and cylindrical vessels. We found the same value for the core density resulted, indicating the extrapolation procedure is appropriate. This best fit value, ϕ_c was taken to represent the infinite system size density.

Finally, we needed to account for loading procedure. For any given shape, a range of packing fractions can result depending on whether or not the aggregate has been mechanically agitated or simply poured in place. We elected to consider packings made both as poured and agitated until reaching an approximately steady state. For agitation we used both tapping by hand and mechanical driving. For hand-agitated systems we tapped the container (typically a spherical flask), until no more particles could be added. For mechanically agitated systems we tapped the container (typically a cylindrical beaker) with approximately 1,000 3 g (g = 9.8 m/s^2) sinusoidal pulses each spaced by half a second. Measurements of the core region packing fraction with these two methods were consistent with one another, suggesting the packings they generate are good proxies for the asymptotic state.

Figure 4.2 shows the packing fractions measured in different sized containers, the best-fit lines for our extrapolation procedure, and, for comparison the simulation

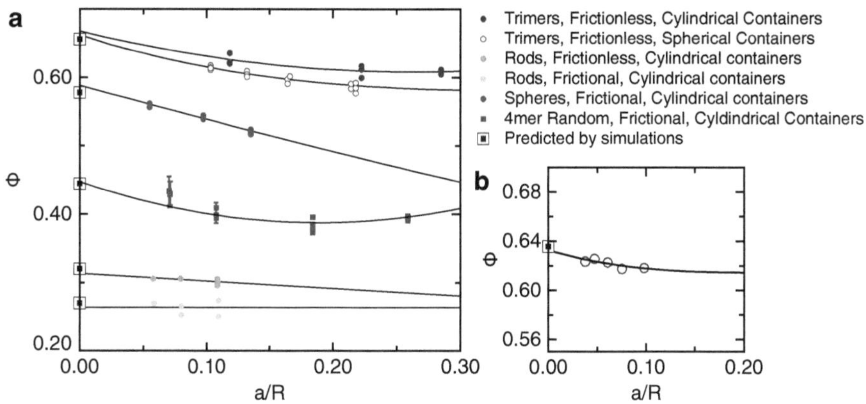

Fig. 4.2 Comparison of packing density from simulations and experiments (Miskin and Jaeger 2014). The plots show the packing fraction ϕ as a function of a/R, where a is the size of the largest constituent sphere in a molecule and R is the container radius. (**a**) Experimental measurements using 3D-printed molecules with lines representing best fits to the model described in the text. *Squares* along the vertical axis $a/R = 0$ are predictions from the DEM simulations. (**b**) Measurements on 3 mm glass spheres in a spherical container, consolidated by tapping. The extrapolated value $\phi = 0.635$ was used as a validation for the simulation procedure and agrees well with literature values, providing further confirmation of the extrapolation method

predictions for the appropriate shape and loading procedure. Note that these data sets were used to generate all the extrapolated packing fractions in the subsequent text.

4.2.2 Simulation Procedure

Our experimental procedure, as defined, required three concepts to be translated into computer simulations. First, the simulations needed to model pouring particles under gravity. Second, the reported value for the packing fraction needed to be independent of system size. Third, we needed simulation parameters that would generate approximations to the as poured packing and the asymptotically tapped packing.

We simulated our packing procedure by settling particles into a box using gravity. The length dimensions in the simulations were set relative to the diameter of a sphere with a volume equal to a single granular molecule. In every simulation, we set 1000 granular molecules into a long, loose column extending 40 diameters from the bottom of the container in which no particles overlap. We chose the height of 40 diameters arbitrarily, although we found no change in packing fraction when we simulated the same procedure with a column of 50 diameters. We then activated gravity causing the particles to settle into a packing. The settling process was stopped when the average distance between particles and the bottom of the box changed by less than 10^{-5} % and the ratio of the average unbalanced force to the average contact force was less than 1 %. We found that lower tolerances did not change the calculated packing fractions.

Once a packing formed, we needed to extract a measurement that represented the infinite system size packing fraction. While, in principle, we could have simulated different sized systems and extrapolated the results, in analogy with our experimental method, we felt this was needlessly costly. Running multiple simulations at different, larger system sizes would significantly increase run time. Furthermore, any single simulation provides us with complete information about particle locations, and thus, gave us a direct way to measure core-region packing fractions. Specifically, we used the particle locations to generate the radical Voroni tessellation for the packing (Rycroft 2009). Then, we made a volume cut by drawing a new, imaginary rectangular box centered on the center of mass. The x, y, and z, dimensions of the box were set equal to the 60th percentile of particle distances measured along each respective axis. We defined the packing fraction by adding up the particle volumes and the Voroni volumes that fall within this box, and dividing the former by the latter.

This simulation method had a number of free model parameters that needed to be appropriately selected to produce good approximations to real packings. As in Chap. 2, our particles interacted via both normal-force linear springs, and frictional springs, with the same default model parameters as discussed there. Following Cundall and Strack (Cundall and Strack 1979), we also included a numerical, global damping to dissipate energy. This numerical damping was configured so that, if two particles collide, 20 % of the energy would be lost after a loading-unloading cycle. When using our simulation protocol to approximate asymptotically tapped packings, we simply set the tangential spring constant to zero. All the simulations were performed using the PFC3D DEM package from the Itasca Consulting Group, (PFC3D 4.0). The code was executed on a small computing cluster made of 10 identical desktop computers.

While there are other possible contact models, dissipation mechanisms or simulation protocols, we accepted this protocol because it produced reliable agreement with our experimental procedures. For validation, we printed spheres and dimers and measured the infinite system size packing fractions with our extrapolation method. We compared the results to those generated by our simulation procedure. For spheres, as poured, we measured a volume fraction of 0.59 ± 0.02. Our simulation protocol calculated a volume fraction of 0.578 ± 0.002, in good agreement. For spheres tapped into the asymptotic state, we measured a volume fraction of 0.63 ± 0.02 while our frictionless simulations measured a volume fraction of 0.635 ± 0.004. Note both of these values are also in good agreement with literature values. For dimers, we measured the as poured volume fraction to be 0.56 ± 0.02 and our frictional simulations calculated a packing fraction of 0.55 ± 0.004.

4.3 Solving the Inverse Problem

4.3.1 Finding the Extreme Packing Shapes

As the first step to answering the 'granular inverse packing problem', we used our evolution framework to identify shapes that generate particularly dense and particularly loose packings when poured under gravity. We allowed our optimizer to make granular molecules from a maximum of 10 constituent spheres of variable radii. Figure 4.3 shows the median fitness for the simulated populations, under our two loading procedures at each generation. After roughly 50 generations, our algorithm converged when the simulations included friction (Fig. 4.3a) and roughly 100 generations when the simulations were frictionless (Fig. 4.3b).

After starting from the random, un-adapted configurations (Fig. 4.3), our algorithm identified three distinct geometries (Fig. 4.4): a sphere, a trimer, and 10mer rod. The trimer shape is symmetric, made from two secondary spheres with radii 0.3 the size of the main sphere, and a bond angle of approximately 70°. These shapes have been adapted to pack densely with friction, densely without friction and loosely, respectively.

To verify the results of the optimizer, we 3D-printed particles and measured their packing fractions under the specified loading protocols. For rods, our simulations predicted packing fractions of 0.271 ± 0.009 and 0.313 ± 0.002 with and without friction. In experiments, we measured packing fractions of 0.26 ± 0.03 and 0.31 ± 0.01 for the as poured packing and the tapped packing, respectively. For spheres, our frictional simulations predicted a packing fraction of 0.578 ± 0.002. Using the

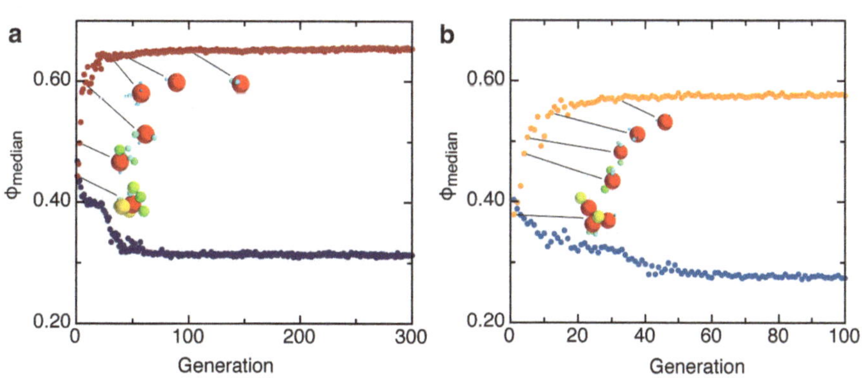

Fig. 4.3 Evolving the densest and loosest packing shapes (Miskin and Jaeger 2014). Starting from random configurations, our evolutionary algorithm adapted particles to pack densely and loosely when poured under gravity without friction (**a**) and with friction (**b**). The median packing densities are plotted against generation number. For comparison, we also include call outs to some of the shapes that generated these median values. Note how our evolution process essentially filters out complicated, but suboptimal shapes to discover surprisingly simple shapes as extreme packers

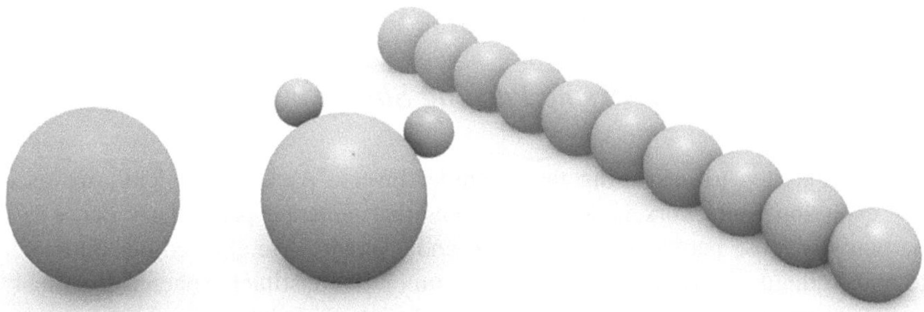

Fig. 4.4 Three extreme packing shapes (Miskin and Jaeger 2014). The optimizer converged to these three shapes as the particle shapes that generated the most extreme packing densities, given our loading protocol and material parameters. While the rod and trimer exhibit several symmetries, so as to appear relatively simple, these are emergent features. They were not programmed as design goals, but rather came about through generations of random mutation and selection

'as poured' experimental procedure we measured a volume fraction of 0.59 ± 0.02, again within error. Finally, we measured a density of 0.67 ± 0.03 for tapped packings of our trimer particle, within error of the simulation prediction, 0.658 ± 0.003.

4.3.2 Analysis of the Results

For the particular optimization goal of finding shapes that minimize or maximize packing density when poured under gravity, the optimization process reduced much of the shape complexity that would have been possible using all 10 spheres. Perhaps surprisingly, all three shapes are relatively simple granular molecules.

Others have identified two of the shapes, rods and spheres, as extreme packing shapes. When considering loose packings prepared by fluidization, Baker and Kudrolli (2010) demonstrated that spheres pack denser than any of the Platonic solids. For rods (Wouterse et al. 2009, 2007; Philipse 1996), it has been reported that the volume fraction for an aggregate can be made arbitrarily low by increasing the number of monomers. The fact that the optimizer used all 10 spheres in constructing the rods suggests that here too the packing fraction could have been made lower had we allowed the optimizer more constituent elements.

Perhaps the most surprising shape identified was the trimer particle. We anticipated cubic or tetrahedral granular molecules to be the densest packing shapes without friction. Yet by directly simulating these shapes with our loading procedure and material parameters, we found them to be suboptimal: cubes and tetrahedra made from equal sized spheres packed to 0.465 and 0.641, respectively.

However, we also note that the trimer particle is not the densest packing particle out of all possible shapes: ellipsoids can pack randomly to a density up to 0.74. We speculate that this discrepancy comes from the fact our optimizer was constrained

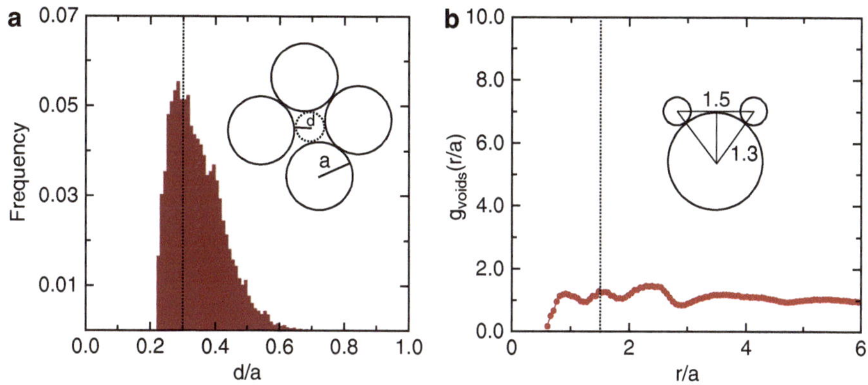

Fig. 4.5 Filling the voids of a sphere packing (Miskin and Jaeger 2014). To rationalize the geometry selected by the optimizer as the densest packing shape, we considered the size (**a**) and pair correlation (**b**) of voids in a random packing of spheres of radius a. Voids have a most probable size of 0.3 a and are distributed roughly 1.5 a apart. The optimizer picked granular molecules consisting of a central sphere with two nubs that nicely fit into these voids (*inset* in **b**)

to build particles from bonded spheres with no overlap. Such shapes always have surface concavities, and as Schreck and O'Hern (Schreck et al. 2010) demonstrated in 2D, there is a pronounced difference in packing fraction between ellipses and dimers comprised of two overlapping, equal-radius disks already when the aspect ratio exceeds 1.2. It would be straightforward to extend the approach outlined here to evolve smoother geometries by introducing a very large overlap between adjacent bonded spheres inside each compound particle; in fact, one could include overlap as a mutable parameter to be optimized.

The optimizer's choice became clear after we examined the void space left in a packing of frictionless spheres. Specifically, we studied points that were equidistant to each particle in a set of nearest neighbors (see sketch in Fig. 4.5a). If a sphere were placed at one of these points and expanded until just touching all the neighbors, it would be as large as possible without requiring the packing to restructure, making each of these sites local maximum for the size of a spherical void. Therefore, designing granular molecules that fill these voids can potentially be a good strategy to maximize packing density, provided the particles are free to explore different local configurations.

To test this hypothesis we took a packing of spheres of radius a, calculated all such local-maximum void sizes and simulated appending spheres to fill them. We skipped any void that overlaps with a void that has been previously added. By making a histogram of the appended sphere radii, d, we found that the most probable void side is $0.3a$ (Fig. 4.5a). This is the same size ratio as in the trimer particle. What's more, we found a surprising structure in the pair correlation function between these voids. Specifically, we found a local maximum in the probability density corresponding to two voids distributed about $1.5a$ apart (Fig. 4.5b). Assuming a void size of $0.3a$ and that these two voids both touch the same sphere, this would correspond a bond angle of 70°. This is the same bond angle as in our trimer. While there is also a second peak at $2.5a$, two $0.3a$ appendages can at most be spaced $2.3a$ apart, making it impossible to exploit this correlation. Evidently, our trimer was designed to exploit features lurking in an ordinary, random packing of spheres.

4.3.3 Creating and Using Design Rules

Combined, the experimental validation and the theoretical rationalization of the shapes found by the optimizer provide a reasonable basis to accept them as good approximations to the true densest and loosest packing shapes. As the next step, we consider how to deform one into the other, and how to use this deformation as a set of design rules.

First, we deform our rod shapes into spheres by sending the sphere radius at the end of the rod to zero, one at a time. We then take the trimer particle and send the secondary spheres to zero radius, again one at a time. These two sets of deformations overlap once the mutated shapes become dimers and are transformed into spheres. Note that, while there are infinitely many ways to perform this deformation, this choice provides a convenient way to index shapes. By systematically sending the furthest spheres to zero, a single number, the sum of all radii in a particle, marks each shape uniquely within the deformation path. For example, if the sum of all radii is 4.2, we know that the shape in question is a linear configuration of four $a = 1$ spheres plus one sphere with $a = 0.2$ sitting at the end.

For each of these mutated shapes, we ran a corresponding DEM simulation, and in Fig. 4.6a, b we plot the resulting packing fractions against our 'shape index,' the sum of the radii in the particle. In Fig. 4.6a, the rods form a lower branch spanning all the packing fractions from 0.65 to 0.3 while the trimers yield a second branch, spanning packing fractions from 0.64 to 0.66. For the frictional simulations in Fig. 4.6b, rods alone span every packing fraction up to 0.59. Thus, every packing fraction between the minimum and maximum discovered by the optimizer can now be mapped to at least one shape. Furthermore, the shape's geometry can be simply read off the graph, given our deformation procedure.

As an illustration of this inverse method in action, we picked an arbitrary packing density, in this case 0.445. Suppose that one wants to find a particle that produces this packing fraction when poured in place. As shown in 4.6b, the intercept between 0.445 and our response curve occurs for a shape with a shape index of about 3.5. Because the response curve was constructed by sending each sphere to zero sequentially, there is only one shape with this value: a linear particle with three identically sized spheres plus a fourth at the end with one half the radius. We 3D-printed this particle and measured its as poured packing density. We found, as desired, a packing density of 0.45 ± 0.01.

One can note that our inversion method is not one-to-one. For example, the shape index vs. packing fraction graph for frictionless packings exhibits non-monotonic

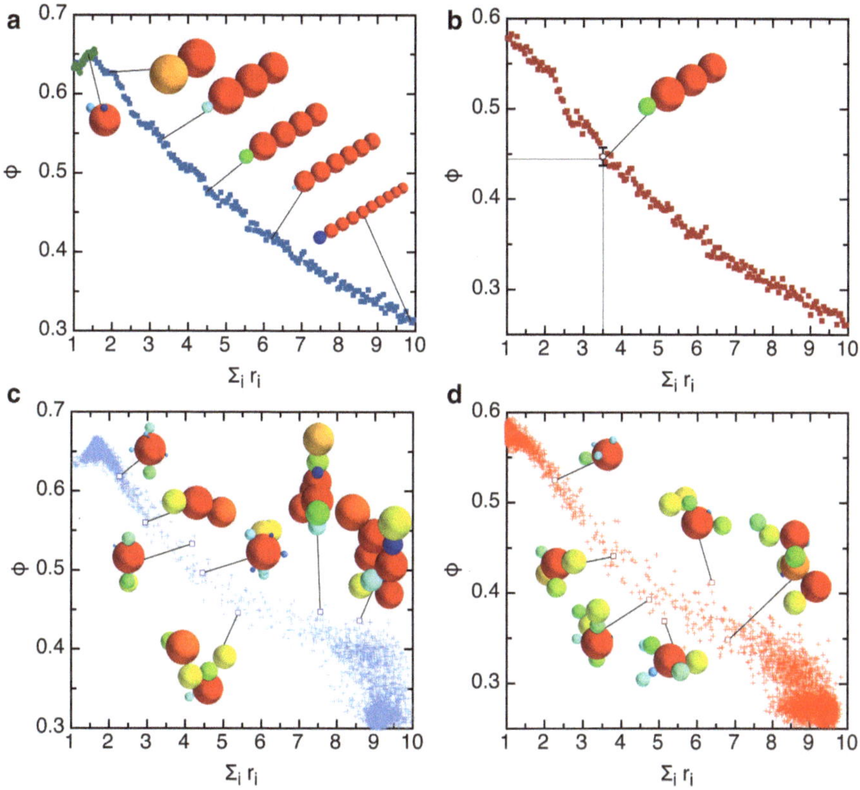

Fig. 4.6 Inverting a packing problem (Miskin and Jaeger 2014). Because the optimizer identified rods, spheres and a trimer as extreme packing shapes, we used these as a basis to build an inverse problem solution. Specifically, we deformed rods (*blue* in **a** and *red* in **b**) and trimers (*green* in **a**), by sending each sphere radius to zero sequentially and simulating the resulting shapes both without (**a**) and with (**b**) friction. This procedure generates the plotted paths that can be used to solve inverse problems graphically (**b**). While this procedure is extremely simple, it is founded on the immense search procedure performed by the optimizer (**c**, **d**). For both simulations with and without friction, the optimizer explored a tremendous number of complex and exotic geometries before ultimately converging on the shapes used as a basis for (**a**) and (**b**). Any one of these exotic behaviors could, a priori, have been optimal, illustrating the powerful simplifications created by our evolutionary method

behavior. Thus, a designer trying to find particles that pack to densities between 0.64 and 0.65 may choose from either an asymmetric dimer or a trimer geometry. From a practical point of view, this degeneracy is unimportant. Both geometries answer the same granular packing problem and so a designer may feel free to choose either one to pack to the desired density.

We contrast simple design rules and the simple shapes they lead to against the other shapes considered by the optimizer while perusing the extrema. In Fig. 4.6c, d we plot all the points it explored in finding the minimum and maximum response shapes with call outs to some of the curious geometries that generated them. It is clear that the optimizer parsed through a tremendous number of exotic and complicated geometries. This large degree of diversity is indicative of how effectively artificial evolution, as a process, filters a broad number of possibilities into an outstanding subset. The optimizer had sufficient evidence to select the shapes in Fig. 4.4 as a basis for design rules only after parsing this wide range of potential configurations.

4.4 Summary

While optimization problems are difficult because only a handful of solutions extremize a function, inverse problems are hard for the opposite reason: there are often a number of objects that meet intermediate functional requirements. The difficulty does not arise from identifying solutions, but rather organizing them. Indeed Fig. 4.6 makes this point clear: the lower panels of c and d show that numerous shapes meet almost any given packing fraction demand. Yet there is no clear way, short of the plotted call outs, to determine what these shapes actually look like. Conversely the upper panels not only link every packing fraction with a shape, but they are deliberately structured to be easy to use.

This is the basic idea that defines design rules and their most powerful advantage: they condense a complicated design problem into a simple set of instructions. Of course design rules, as envisioned here using shape, are not limited to this particular problem. Had we considered packings formed by a protocol different from simple pouring, for example isotropic compression, we may have found different optimal shapes and different maximum/minimum volume fractions. As a case in point in this chapter, the inclusion of friction produced different shapes as densest packers. Yet the optimizer operates independently from the simulation procedure at hand, and so these cases would be straightforward generalizations.

Further, as a first order approximation, the capacity to rationally select shapes that produce desired packing fractions implies the capacity to tune fluid and gas flow through the interstices of a granular bed (Carman 1997; Malinouskaya et al. 2009; Carman and Carman 1956), thermal and electrical conductivity (Batchelor and Obrien 1977; Malinouskaya et al. 2009), diffusivity (Carman and Carman 1956), and a host of other physical properties. However, the relationship between shape and packing fraction explored here serves simply as a proof of principle; any other relationship between particle shape and one or more aggregate properties of the packing could be posed as a granular inverse problem, and solved over some range of values by this method.

Moving beyond shape in granular materials, we speculate that similar design concepts will work well on other systems. Abstractly, our workflow is to identify a control parameter and optimize the material response with respect to it. Once this is achieved, a user must construct some series of small distortions that deform the minimum producing configuration to the maximum configuration. By selecting this path intelligently, the design problem can be reduced to one dimension. And, in one dimension, the results become manageable by brute force and can be plotted in a simple graph. The only caveat is that the deformation mechanism must produce a continuous variation in responses between the min and the max, which to us does not seem overly restrictive for material design problems.

The fact that complex, open-ended inverse problems can be reduced to a graph exercise speaks loudly to the power in automatic materials design. In a grand scope, optimization is simply a starting point: it provides a well defined bridge from macroscopic goals to microscopic features. But once this bridge is built, the task becomes to expand on the results and extract new functionalities. The concepts that do this, like design rules, then become the integral steps for moving away from exploration and towards true design.

References

Baker, J., & Kudrolli, A. (2010). Maximum and minimum stable random packings of platonic solids. *Physical Review E, 82*(6), 061304.

Batchelor, G. K., & Obrien, R. W. (1977). Thermal or electrical-conduction through a granular material. *Proceedings of the Royal Society of London Series A-Mathematical Physical and Engineering Sciences, 355*(1682), 313–333.

Carman, P. C. (1997). Fluid flow through granular beds. *Chemical Engineering Research and Design, 75*, S32–S48.

Carman, P. C., & Carman, P. C. (1956). *Flow of gases through porous media*. London: Butterworths Scientific Publications.

Cundall, P. A., & Strack, O. D. (1979). A discrete numerical model for granular assemblies. *Geotechnique, 29*(1), 47–65.

Donev, A., Cisse, I., Sachs, D., Variano, E., Stillinger, F. H., Connelly, R., Torquato, S., & Chaikin, P. M. (2004). Improving the density of jammed disordered packings using ellipsoids. *Science, 303*(5660), 990–993.

Jaoshvili, A., Esakia, A., Porrati, M., Chaikin, P. M. (2010). Experiments on the random packing of tetrahedral dice. *Physical Review Letters, 104*(18), 185501.

Malinouskaya, I., Mourzenko, V. V., Thovert, J. F., & Adler, P. M. (2009). Random packings of spiky particles: Geometry and transport properties. *Physical Review E, 80*(1), 011304.

Miskin, M. Z., & Jaeger, H. M. (2014). Evolving design rules for the inverse granular packing problem. *Soft Matter, 10*(21), 3708–3715.

Philipse, A. P. (1996). The random contact equation and its implications for (colloidal) rods in packings, suspensions, and anisotropic powders. *Langmuir, 12*(24), 5971–5971; *12*, 1127.

Rycroft, C. H. (2009). Voro++: A three-dimensional voronoi cell library in c++. *Chaos, 19*(4), 041111.

Schreck, C. F., Xu, N., & O'Hern, C. S. (2010). A comparison of jamming behavior in systems composed of dimer- and ellipse-shaped particles. *Soft Matter, 6*(13), 2960–2969.

Suzuki, M., Shinmura, T., Iimura, K., & Hirota, M. (2008). Study of the wall effect on particle packing structure using x-ray micro computed tomography. *Advanced Powder Technology, 19*(2), 183–195.

Wouterse, A., Luding, S., & Philipse, A. P. (2009). On contact numbers in random rod packings. *Granular Matter, 11*(3), 169–177.

Wouterse, A., Williams, S. R., & Philipse, A. P. (2007). Effect of particle shape on the density and microstructure of random packings. *Journal of Physics-Condensed Matter, 19*(40), 406215.

Chapter 5
Transitions of Designs

Abstract A new host of materials science problems open up once the task of identifying extreme materials has been automized. This is because optimization condenses the wide range of possible material microstructures into only those that emphasize extraordinary behaviors. In this chapter we ask how materials that have been optimized for a particular task change when the parameters that define the optimization are varied. Specifically, we consider which particle shape cools the fastest in a granular gas. We show that the answer depends crucially on the density and number of elements in the shape and that transitions between classes of optimal shapes take place as these parameters are varied. We use these transitions to isolate new pieces of physics that describe the cooling processes of the gas. More generally, we show how material optimization enables a new workflow to isolate the relevant pieces of physics that govern a material's bulk behavior.

5.1 Introduction

5.1.1 Stability of Designs

The results of an optimization are to varying degree dependent upon the simulation paramcters used. For example, in Chaps. 3 and 4 we found that the trimer shapes returned as the stiffest and densest packing, respectively, are dependent on the magnitude of frictional stiffness. However in Chap. 3, the variations were quantitative, but not qualitative: 3-mer triangular particles remained the stiffest shape whatever the value for the frictional stiffness. Conversely in Chap. 4, the qualitative structure of the densest packing shape depended on whether friction was included. Indeed, the inclusion or removal of friction produces particles with different numbers of constituent spheres.

These features are reminiscent of bifurcation theory. For a small change in the simulation parameters, it is possible that only the steepness of the landscape changes, while the minima location is essentially persevered. In this case, the result of the optimization will be unaltered, as in the case of particle stiffness. However, there is also the possibility of a bifurcation, accompanied by the appearance of new, different minima, as in the case of particle packing.

M.Z. Miskin, *The Automated Design of Materials Far From Equilibrium*,
Springer Theses, DOI 10.1007/978-3-319-24621-5_5

By analyzing where or how minima change, we can hope to gain important insight about the underlying physics. One means to this end is a kind of bifurcation diagram, in which the minimized variables are plotted against the system parameters. By observing where qualitative changes onset, one can distinguish particular simulation settings as important balancing points between regimes with different physics.

In this chapter we borrow concepts from bifurcation analysis and apply them to design problems. The basic program is to repeatedly optimize a macroscopic property for every simulation parameter that falls in a grid. Then, we plot the results in the grid to search for qualitative differences. In some regimes, we expect the optimization results to be robust, and that small changes in parameters lead to little or no change in the optimized object. In other cases, we expect phase transitions may occur, under which the optimization result differs significantly over small variations in the simulation parameters. By deciding upon a boundary dividing where these events happen, we construct a phase diagram for designs.

5.1.2 Designing a Granular Fluid to Freeze

To strengthen the analogy with phase transitions, the system we consider is a "granular fluid," generalized to include shape. Specifically, we study a collection of bonded spherical particles, flying freely in space, that loose energy upon collision. The bonds between the spheres in a particle are flexible and resist compression, shear and bending. By this construction, this fluid is composed of two subsystems, translational motion of the particles and the internal degrees of freedom in vibration. Left undisturbed these systems evolve conservatively. However when energy is transferred between the two via a collision, a small loss of total energy also occurs. Prior work has studied a similar system in 1-D by using rods with vibration modes as the internal degrees of freedom (Giese and Zippelius 1996; Aspelmeier et al. 1998; Aspelmeier and Zippelius 2000). In effect, we attempt to generalize the results to 3-D by agglomerating the rods into composite particles.

Naturally, in the long time limit the dissipative gas must decay to a zero translational energy configuration. However, we also expect that gasses made from some shapes will decay faster than others. For example, if the fluid is dilute, shapes with a large cross sectional areas will collide more frequently, and thus experience more opportunities to dissipate energy. Alternatively, if the fluid is densified, then some shapes may jam, producing large amounts of dissipation immediately. Further, the modes of vibration are also determined by particle shape. If the modes are conducive to increasing the number of collisions, then the shape could dissipate more. Conversely, if after a collision the two particles are pushed apart by vibrations, the gas may dissipate less.

These speculations suggest that whether a particle shape looses energy quickly or slowly will depend significantly on the initial density of particles, ϕ, as well as the number of elements in the shape, n. To examine these particular relationships, we seek out all the optimized shapes that cool fastest made from $n = [3, 4, 5, 6]$

particles and starting at an initial density of $\phi = [0.05, 0.10, 0.15, 0.20, 0.25]$. For each ϕ and n, we optimize for dissipation by finding the particle that dissipates the most energy over a long, fixed time interval. When plotted together, we find that the optimized shapes come in three broad families: arms, rods, and compact objects. Moreover, we build an intuition about new physics that causes linear shapes to transition into compact ones.

5.2 Simulation Methodology

The granular fluid is simulated by molecular dynamics. Each particle is constrained to evolve in a box of periodic boundary conditions, 10 particle diameters in length. The particles are made of bonded spheres which are scaled so that the total particle mass is fixed to $\frac{4}{3}\pi(0.5)^3$ in every simulation. The granular molecules are seeded in random positions and orientations within the box. At the start of each simulation, a random center of mass motion is given to each particle as an initial condition so that each coordinate velocity is normally distributed. The total center of mass momentum is then calculated and subtracted from all the particles to put the measurements in the center of mass frame.

During a collision, two particles interact by normal spring forces with a spring constant set to 10. Further, we include an underdamped viscous dashpot on the normal springs to create an effective coefficient of restitution, fixed in all the simulations to 0.8.

The bonds between each sphere in the composite particle are flexible and resist compression, shear and bending. We model each bond as a simple elastic beam with an elastic modulus set 100 times the spring normal stiffness. This choice puts us in the limit where the internal degrees of freedom oscillate much faster than the collision duration, so that the internal degrees of freedom interact frequently with the translational modes, even in the dilute limit.

Finally the system is left to evolve for 25 units of time. For a corresponding ideal gas of equal volume spheres, this would produce roughly 1,000 total collisions per run. Though we picked this total duration arbitrarily, we found that running the simulation for a longer period of time does not seem to change the order shapes appear in a list ranked by remaining energy. Once the ranking results are static, then as far as the optimization process is concerned, running the simulation for longer is superfluous.

5.3 Examining Transitions Between Designs

5.3.1 Finding the First Shapes to Freeze

To get a broad sense of how shape impacts the cooling dynamics, and how the results relate to the underlying physics, we considered particles made of 3,

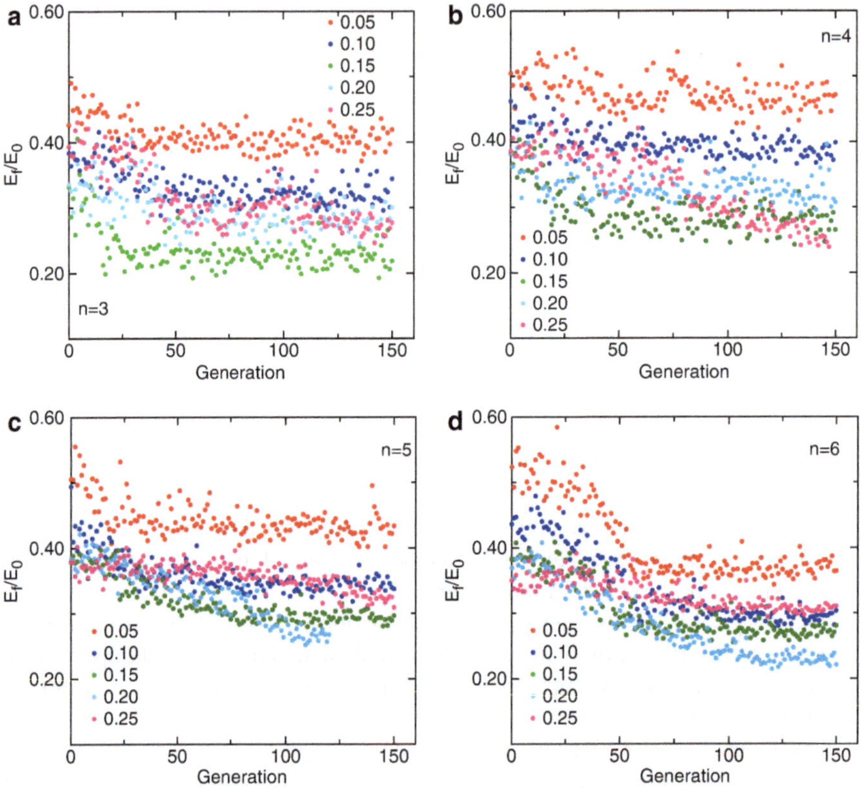

Fig. 5.1 Median values of the final energy in each generation for shapes made of 3 (**a**), 4 (**b**) 5 (**c**) and 6 (**d**) particles at various volume fractions. Notice how the median energy values do not decrease monotonically with volume fraction. Instead, they begin to increase at a volume fraction of around 0.20. This increase corresponds to a qualitative change in the optimized geometries from linear shapes to compact ones

4, 5, and 6 constituent elements when initialized to packing fractions of $\phi = 0.05, 0.1, 0.15, 0.2,$ and 0.25. Since we are interested in qualitative rather than quantitative features, we reduced the run time of each optimization to 150 generations. Grand total, we ran roughly 24,000 simulations to discover 20 optimized geometries.

Plotting the median values per generation in Fig. 5.1 shows that convergence occurs in the allotted 150 generations for all 20 optimizations. For low densities, we note that the median actually changes very little as the optimization proceeds for all four particle numbers. Furthermore, increasing the number of elements in shape, causes particle shape to play a more dominate role in dissipation.

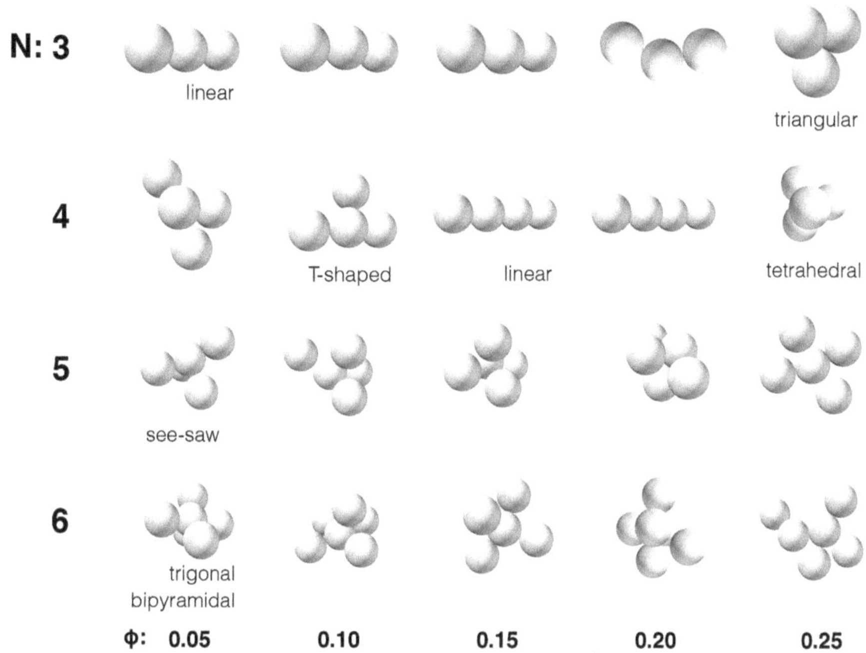

Fig. 5.2 The geometries identified as first to freeze at different volume fractions ϕ and made from different numbers of elements n. While varying ϕ, some optimization results are stable while others give way to qualitatively different regimes. Conversely examine shapes of varying n shows how motifs emerge linking families of shapes together: at high volume fraction there are compact shapes, at intermediate values rods appear, and at low volume fraction shapes with arms surrounding a central sphere are dominant

5.3.2 *Families of Fastest Cooling Shapes*

By plotting the optimized shapes in a grid (Fig. 5.2), we immediately see families of shapes and transitions in design stability. For trimer particles, we find that rods act as the most dissipative shape provided the initial packing density is less than roughly 20 %. Above this critical value, a compact triangular geometry becomes the fastest cooling shape. When the number of monomers in the particle is increased to 4, a new class of particle shapes appears at low densities, given by a "T-shaped" geometry. These shapes are stable up to a packing fraction of 0.15, at which rods reappear as the most dissipative shape. At a packing fraction of 0.2 for the 4mer shapes, a transition again occurs where rods get replaced by tetrahedra. For particles made from 5 and 6 elements, we found particles with arms are a stable solution over the full range of packing fractions, although the exact shape appears to change at high densities.

When going to larger numbers of spheres at a fixed volume fraction, a motif typically emerges, much like in Chap. 3. For example, it seems as though an armed particle made of n spheres could be generated by simply appending a sphere to its $n - 1$ counterpart. We note that virtually all of the shapes with bear an incredible resemblance to the molecular geometries considered in valence-shell electron pair repulsion (VESPR) theory, and we have adopted some of the corresponding names and labeled particularly striking examples. Furthermore, several of the shapes that result at low packing fraction correspond to VESPR models with a steric number of 5: they could all be formed by arranging 5 spheres to be maximally distant from one another on a shell surrounding the central particle then removing elements.

5.3.3 The Transition from Linear to Compact Geometries

Altogether, we find a total of 3 basic particle families: particles with arms, rods, and compact geometries, with two transitions in the range of densities examined. One of these transitions can be elucidated by examining the total energy in the system, as well as how it is distributed. Figure 5.3 shows how the energy for each granular fluid decays over time. For most of the curves the energy values at a given time decreases monotonically with increasing density. But for the 3mer and 4mer particles plotted in Fig. 5.3a, b, we notice that the shapes at the transition between linear and compact geometries ($\phi = 0.20$) typically retain more energy than the shapes at lower factions. We note that this change is also reflected in the median values plotted against generation in Fig. 5.1, implicating that is a generic feature for all shapes simulated at these volume fractions and not just an error on the part of the optimizer.

At first glance this result is counterintuitive: one would think that increasing the fluid density would increase the collision rate, and so a granular fluid should always loose more energy at higher densities. But by analyzing the amount of energy trapped in the vibrational modes, plotted in Fig. 5.4, the underlying physics become clear. For the low density fluids, energy slowly builds up in the vibrational modes, until reaching a roughly constant value. As the density increases, a peak begins to appear in the energy stored in vibration. The peak for n = 3 and n = 4 becomes the most pronounced at $\phi = 0.20$ when it stores roughly 20 % of the total energy early in cooling process. At largest volume fraction $\phi = 0.25$, however, we find the size of the peak is again reduced.

If extra energy is stored in the vibrational modes, the particles must be moving slower in the fluid, thereby colliding less frequently. Thus, if energy is efficiently transferred to vibration, a large initial number of collisions will be increasingly difficult to sustain, thereby decreasing the total rate of energy loss. One could rationalize the transition from linear to compact shapes is a move to mitigate early energy transfer into vibrational modes, facilitating more collisions over the full duration of the cooling process.

Plotting the energy stored in spring-spring interactions between two colliding particles versus time (Fig. 5.5) gives a clearer picture about how energy finds its way into the vibrational modes as the density is increased. For every curve, regardless of the density or number of constituent particles, there is a peak in the stored collision

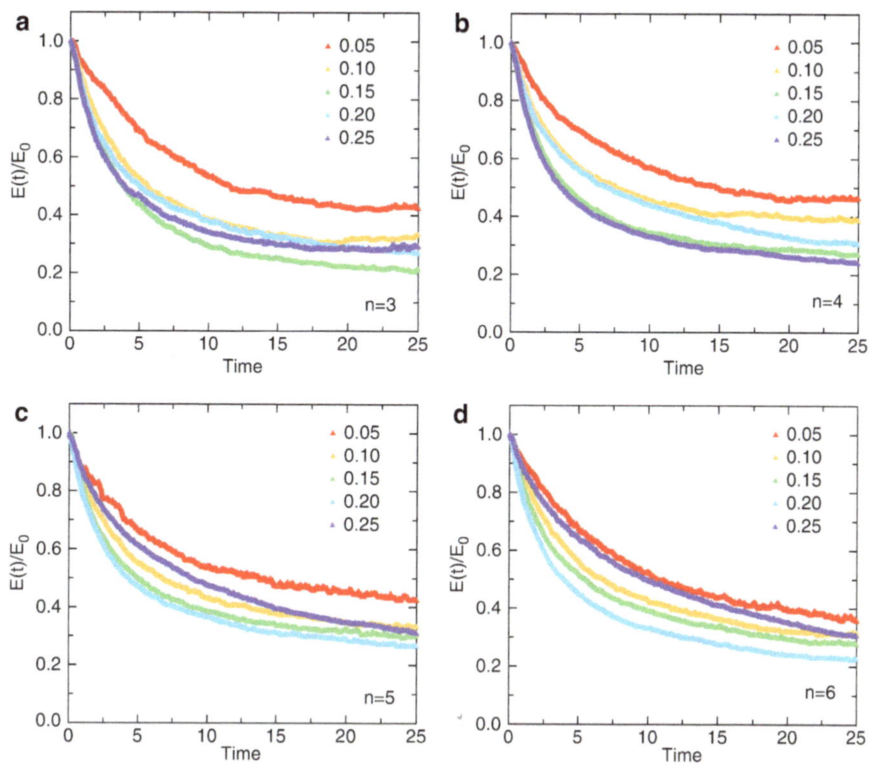

Fig. 5.3 An ensemble average over five runs of the system energy as a function of time for granular fluids of optimized particles at different volume fractions with 3 (**a**), 4 (**b**), 5(**c**), and 6 (**d**) constituent spheres

energy followed by a decay, lasting roughly 0.8 units of time. We note that the decay time for the peak is approximately the period for the simple harmonic repulsion that characterizes the particle-particle interactions. However, we also note that the peak height increases monotonically with the initial fluid density, ϕ. If we estimate the mean time between collisions as $\tau_c = \frac{R}{3\phi v_0}$ where R is the particle size and v_0 is the initial velocity, then we find that for every volume faction greater than 0.05, the collision time is shorter than the duration that two particles are in contact. Thus we expect that early on, a large number collisions involving multiple objects take place.

We note that for a highly compact object, any given constituent sphere is bonded multiple times to several nearest neighbors. Effectively, this means that the constituent particles are stiffer than those in rods. As a consequence, when storing a fixed amount of energy in the total particle under compression, more of it will reside in the particle-particle repulsion springs than the bonds, at least when compared to rods. Since for high densities, an increasingly large amount of the total energy gets stored in compression, it makes sense to stiffen the particles, so that when the collision is complete, as much energy as possible remains in the translational modes to maintain the collision process.

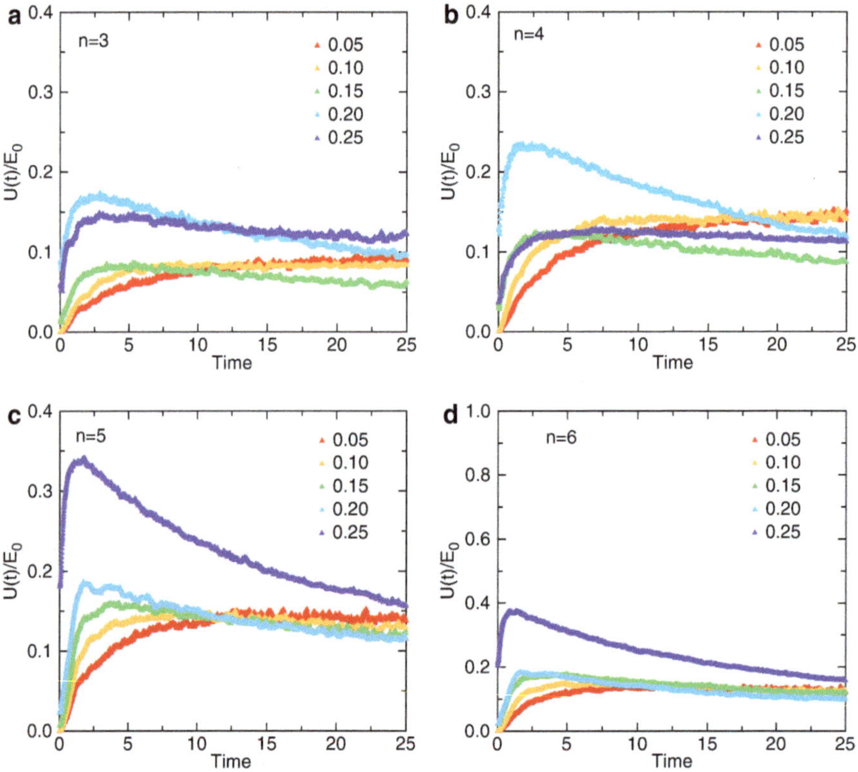

Fig. 5.4 The percentage of the initial energy stored in vibrational modes, $U(t)$ as a function of time. For 3mer and 4mer particles, we note an increase in energy storage occurs at the volume fraction just before the designs change from rods to compact geometries

5.4 Summary

The model for a cooling granular gas studied here is fairly complex. Objects can vibrate, collide, loose energy and have relevant structures. At first glance, separating out what matters seems like daunting task. A priori it is unclear what effect, if any, internal modes have at low densities. If they can be ignored, at which densities does this assumption break down?

By designing objects in the gas for a particular purpose, we gained some hints to move foreword on these questions. Because objects are optimized by leveraging particular physical effects, qualitative changes in design often correlate strongly with qualitative changes in the relevant physical picture. Indeed here we showed that optimized geometries change from the least compact shapes to the most compact shapes at a special volume faction around 20 %. This transition is apparently associated with internal vibrations taking a new role in the cooling dynamics, by prematurely cooling the gas.

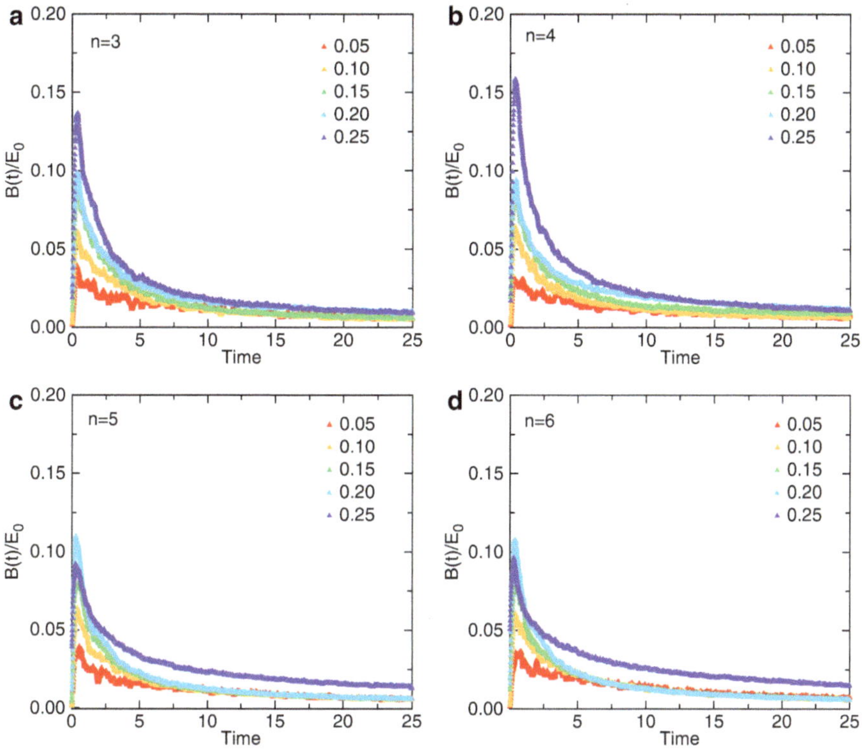

Fig. 5.5 The percentage of the initial energy stored in repulsive springs, $B(t)$ as a function of time. In general, as the volume fraction of a granular fluid is increased, the maximum percentage of energy ever stored by springs during collisions increases. Further, the decay time associated with the relaxing energy stored by these springs in compression is essentially fixed by the spring constant and the particle mass. Thus the width of the peak is roughly independent of volume fraction

Moreover, we also see how motifs emerge as a consequence when the relevant physics stays the same. By examining the similarities between shapes, we gain clues about what effect they are trying to produce. This again allows us to make guesses about what physics is behind the result.

Juxtaposing these similarities and differences, as in Fig. 5.2, arguably provides the best tool to put optimizations in context. In principe, one is able to put each type of result into a different domain of behavior and examine where and why different physical pictures emerge. In this sense both this chapter and the prior, show how to generalize the results from optimization. Yet whereas design rules generalize optimizations to broader applications, here we worked to expound new physics from the optimized states.

References

Aspelmeier, T., Giese, G., & Zippelius, A. (1998). Cooling dynamics of a dilute gas of inelastic rods: A many particle simulation. *Physical Review E, 57*(1), 857.

Aspelmeier, T., & Zippelius, A. (2000). Dynamics of a one-dimensional granular gas with a stochastic coefficient of restitution. *Physica A: Statistical Mechanics and Its Applications, 282*(3), 450–474.

Giese, G., & Zippelius, A. (1996). Collision properties of one-dimensional granular particles with internal degrees of freedom. *Physical Review E, 54*(5), 4828.

Chapter 6
Online Design

Abstract In this chapter, we explore the possibility of building optimization algorithms for the specific purpose of solving materials design problems by integrating them with statistical physics. Specifically, we construct a formalism that can be used to turn statistical physics models that describe materials into optimizers that tailor them. On two simple test problems, we show that the resulting algorithms can be fast and efficient: we use our framework to trap a particle randomly walking on a substrate and to find optimal interaction energies that cause a simply polymer model to fold into a target shape. The speed of our approach suggests that optimizers developed by our new formalism might fill a niche in tailoring materials where computational expensive simulations prohibit alternative, established methods.

6.1 Introduction

In Chap. 3, we made the claim that a material, at the micro-scale, can only be tuned by the shapes, interactions and histories of its constituents. While the subsequent chapters had all focused on shape as a control parameter, we now turn briefly to designing interactions.

Already, there are litany of approaches that appear in diverse fields: inverse statistical mechanics (Torquato 2009; Jain et al. 2014), protein design (Dahiyat and Mayo 1997; Kuhlman et al. 2003), and crystal structure design (Oganov et al. 2011), to name a few. Often, these approaches divide the labor between an optimizer and a forward simulator, as in the preceding chapters. The optimizer makes a series of guesses and each is independently evaluated by the forward simulator. Consequently most of the computational effort comes from the simulation engine. Furthermore, most of the information produced by the simulation is ultimately discarded when a single number that defines solution quality is returned to the optimizer.

While these approaches work exceedingly well, the clean split between optimizer and simulator begs the question of whether or not additional information generated by the simulation procedure could be built into the optimization scheme. For example, suppose we wanted to design the strengths of pairwise potentials so a thermalized system assumes a particular structure. Then the forward simulator would have to generate samples distributed by the canonical ensemble, while the fitness function might average over how well each sample conforms to the design

M.Z. Miskin, *The Automated Design of Materials Far From Equilibrium*,
Springer Theses, DOI 10.1007/978-3-319-24621-5_6

goal. In doing so, we completely discard any information about how fluctuations in solution quality correlate with fluctuations in interaction energy. In theory, this represents unused data that could inform updates to the interaction parameters in conjunction with the average quality.

In this chapter, we explore the possibility of building an algorithm that integrates directly with the simulation engine to set material interaction parameters. Specifically, we seek an algorithm that uses information obtained online, from an active simulation, to decide how to make small changes to the simulation settings. This means that the flow of information must move both ways between optimizer and simulation engine. Furthermore, we seek an optimizer that uses as much information as possible in the simulations to proceed. Ideally, the optimizer should interpret simulations as inherently statistical objects, rather than trying to force them to behave as simple deterministic fitness functions.

Phrasing the problem in terms of distributions suggests a way foreword: manipulating distribution parameters to make desirable configurations appear more frequently is the motivation behind evolutionary algorithms, as discussed in Chap. 2. It stands to reason that the same procedure used there can be extended to a physical distribution, like the canonical ensemble or even distributions outside of equilibrium. Analogous to how the CMA-ES works to optimize quantities by picking parameters for Gaussian distributions, these algorithms will work to optimize user specified, material properties by picking the physical parameters in the relevant distribution.

Towards these goals, we consider two simple problems: a particle trapped on a substrate and a toy model protein. In both cases, we derive the evolutionary algorithm that adapts to the appropriate physics. In the case of a particle on a substrate, the algorithm is tasked with placing a particle in a particular energy minima by adjusting an electric field and the system temperature. In the case of the model protein, the algorithm must set the interactions in a chain of 6 particles so they fold into an octahedron. These results represent the first steps towards online design. Further, they suggest that this framework may find a niche in optimizing complicated, black-box models where expensive simulations prohibit alternative, established methods.

6.2 Algorithmic Framework

In Chap. 2, we generalized the update equations for evolutionary algorithms to work on arbitrary probability distributions. Consequently, we can use this formalism to generate an optimization algorithm that operates on any object that is statistical in nature (Ollivier et al. 2011). The only conditions are that we must be able to express the structure of the probability density function and sample from it.

These conditions can be fulfilled for physical systems by turning to statistical mechanics. Furthermore, for many physical systems, the probability density function is in the exponential family, a case that was worked through in detail in Chap. 2.

As definite examples of exponential family distributions, we consider systems in thermal equilibrium described by Hamiltonians H that can be broke up into sub-interactions $H/kT = \sum \lambda_i h_i(x)$. Although we could use the evolution formalism from to tune every parameter in the Hamiltonian, for simplicity, we only consider the relative coupling strength compared to kT, denoted by λ_i. For a given design task, we seek a collection of λ_i that extremizes the average value of some goal function, $g(x)$. To evolve the probability distribution to promote that goal, we follow the derivation in Chap. 2 and adapt the parameters λ_i, to follow Eq. 2.19:

$$\dot{\lambda}_i = Cov[h_i, h_j]^{-1} Cov[h_j, f(x)] \tag{6.1}$$

where $f(x)$ is the rank based weighting scheme that promotes configurations that minimize the design goal $g(x)$.

In this new, integrated approach, we only run one simulation per generation to generate samples from the distribution $\exp[-\sum \lambda_i h_i(x)]/Z$. Samples from this distribution at different points in configuration space, x, correspond to the evolved objects, and correlations between the interaction energies h_i and the fitness function drive the updates to the coupling constants λ_i. The process is iterated to produce a flow of interaction strengths, that attempts to improves the average solution quality at each step. As desired, the optimizer analyzes fluctuations, as well as fitness, in picking the next set of simulation parameters. Furthermore, we note that the energy functions h_i do not have to be explicitly known or differentiable. They can be left black-box since they only appear in expectation values. Figuratively, we can say that the algorithm is tasked with empirically learning statistical mechanics in order to design interaction strengths.

Provided one can sample from the appropriate distribution, the only remaining steps become calculating the appropriate covariances, producing a matrix inversion, and numerically integrating the equations of motion for the parameters. Following the CMA-ES derivation, we perform this last task with an Euler approximation.

Finally, we comment that the update rules only act on as many parameters as the user can tune. If, for example, some component of the system's Hamiltonian simply cannot be changed in the lab, then we leave this component out of the update equations. Thus the optimization scheme can be tailored both to the problem definition and to any constraints on experimental implementation.

6.3 A Particle on a Substrate

6.3.1 *Problem Definition*

We start with a particularly simple example: a Brownian particle trapped on a 2D substrate in equilibrium. The substrate applies a potential to the particle, making some positions more likely than others. The optimizer's task is to trap the particle in a specific well. To do so, it is afforded a set of control knobs that morph the energy landscape the walking particle sees. Thus, the complete task for the optimizer is to learn how to set each control knob to make the desired minimum the global minimum, and then quench the system temperature.

6.3.2 Simulation Methodology

Given a set of interaction strengths λ_i, we generate samples for the particle position by running a Metropolis-Hastings random walk. The proposals are generated by adding Gaussian noise to each particle position and the proposal length scale is set equal to $l_p/\sqrt{max[\lambda]}$ where l_p is the characteristic length in the Hamiltonian (we will note l_p explicitly in the following examples). We discard the first 10,000 steps of the sampling algorithm and then allow it iterate 50,000 times. The positions at each of the 50,000 steps are then thinned down to 50 by taking one sample every 1,000 iterations and these 50 points are sent to the optimizer.

6.3.3 Controlling the Particle with a Linear Field

To make things concrete, we first give the optimizer the freedom to tune the interaction strength with the substrate, and control over a linear electric field that it can use to drive the particle. We choose the substrate potential $h_s = \sum_i -\cos[x_i \frac{2\pi}{5}] + x_i^2/25$ and with the field included the total Hamiltonian becomes

$$H = \lambda_s h_s + \lambda_{x_1} x_1 + \lambda_{x_2} x_2 \tag{6.2}$$

$$H = \lambda_s (\sum_i -\cos[x_i \frac{2\pi}{5}] + x_i^2/25) + \lambda_{x_1} x_1 + \lambda_{x_2} x_2 \tag{6.3}$$

where the λ terms are the control parameters for the optimizer. We take the problem length as $l_p = 5$.

Given this form, it should be a straightforward task to tilt the potential to make any design target the global minimum. Once this is done, all the optimizer has to do is cool the system to zero temperature to complete the optimization. For definiteness, we say that the target well is the point $(5, 5)$.

At this stage, we plug in the appropriate covariances into the evolution Eq. 2.19. Doing so explicitly, we find that at each iteration, the algorithm should adjust the interaction strengths according to

$$\frac{d}{dt}\begin{bmatrix}\lambda_s\\ \lambda_{x_1}\\ \lambda_{x_2}\end{bmatrix} = \begin{bmatrix}Cov[h_s,h_s] & Cov[h_s,x_1] & Cov[h_s,x_2]\\ Cov[h_s,x_1] & Cov[x_1,x_1] & Cov[x_1,x_2]\\ Cov[h_s,x_2] & Cov[x_2,x_1] & Cov[x_2,x_2]\end{bmatrix}^{-1}\begin{bmatrix}Cov[h_s,f]\\ Cov[x_1,f]\\ Cov[x_2,f]\end{bmatrix} \tag{6.4}$$

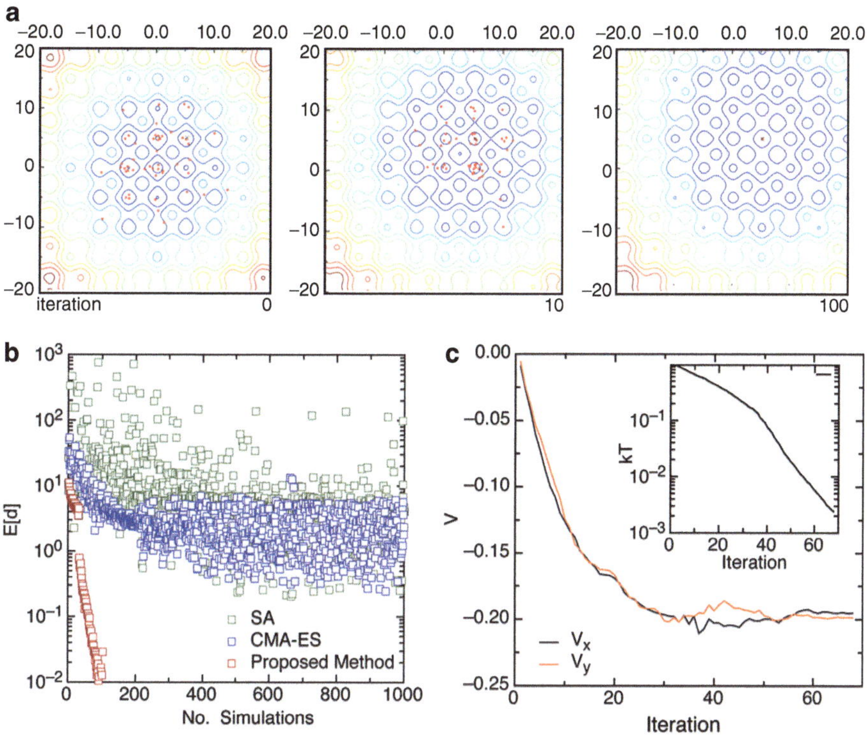

Fig. 6.1 Energy landscapes developed by the optimizer as it worked to trap the particle in at the point (5,5), marked by a *black* x in the figure. *Red dots* correspond to points drawn from a canonical ensemble given the energy landscape

which we integrate with an Euler approximation and a small time step of 0.5. Further we initialize the algorithm with the substrate interaction set to 1 and the field parameters set to zero.

In Fig. 6.1a we plot the energy landscapes generated by the optimizer, as well as the points sampled during each generation. It shows the optimizer quickly learns to tilt the landscape, correctly making the target well the global minimum. The algorithm then spends the remaining time cooling the ensemble, starting after 40 generations (Fig. 6.2b). In short, the optimizer correctly reproduces the anticipated behavior.

The point here is not that the optimizer discovered the optimum parameter settings: the answer was known from the start. Rather, the interesting result is that it was able to do this while the problem was specified so abstractly. Notice that we never explicitly told the optimizer how each energy appearing in the update equations was calculated. We never differentiated energies or fitnesses. We didn't explain to the optimizer, a priori, which knob was temperature and which corresponds to the field. Instead, the optimizer used black-box, derivative

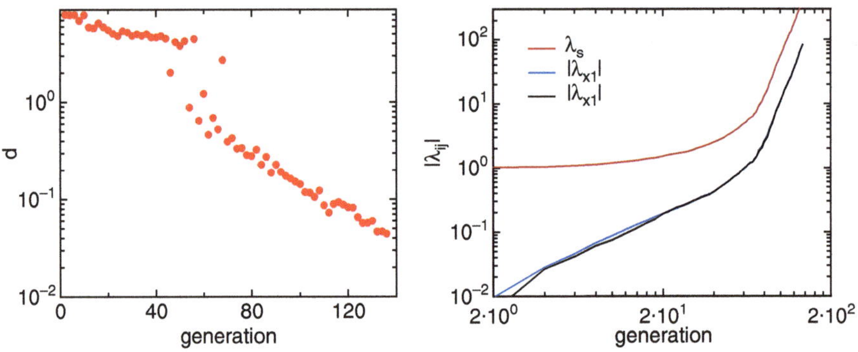

Fig. 6.2 (**a**) Median distance to the target point at (5,5) vs. generation number for a optimizer tuning an energy landscape using temperature and a linear field. Around 50 generations, there is a discontinuity in the slope of convergence. At this generation, the algorithm begins cooling the system as seen by the increase in λ_s plotted in (**b**)

free updates to learn what the parameters represented and to control them to solve a problem. We emphasize this point by moving to a slightly more complicated version of this design task.

6.3.4 Controlling the Particle with a Complicated Potential

If the optimization engine truly learns to adapt interaction strengths on the fly, is should be possible to change just three lines of code, the lines that define each energy contribution, and rerun the exact same algorithm to solve a different, more complicated problem. Following this line of argument, we take the exact same evolutionary algorithm and arbitrarily change the energy landscape to the form

$$H = \lambda_0 J_0(\frac{2\pi}{5}r) - \lambda_1 J_1(\frac{2\pi}{5}r)\cos[\theta] - \lambda_2 J_1(\frac{2\pi}{5}r)\sin[\theta] \tag{6.5}$$

where J_n is a Bessel function of the first kind, order n and $r = \sqrt{x_1^2 + x_2^2}$ and $\theta = \arctan[x_1/x_2]$. Again, we ask the optimizer to place the particle at the point $(5, 5)$. Yet now the choice for each λ is far less obvious because the effects of each control knob are entangled: raising the value of one knob doesn't correspond to a simple translation in one coordinate.

After initializing the algorithm with each $\lambda = 1$, we ran the optimization code. The energy landscapes at a sample generations are plotted with the points sampled in Fig. 6.3. Initially, the point $(5, 5)$ is a maxima in energy. Yet the algorithm has no difficulty adjusting the interaction parameters to place a new, global minimum at the point $(5, 5)$ and then cools the system. Indeed in plotting the median fitness value against generation number (Fig. 6.4), we find an exponential convergence, remarkably similar to the case of the CMA-ES.

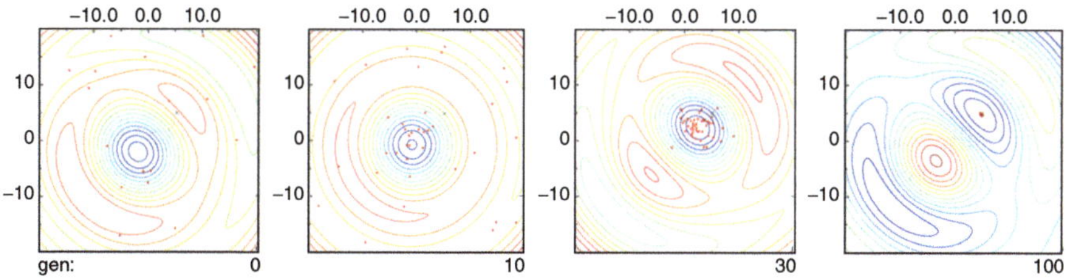

Fig. 6.3 Energy landscapes created by the optimizer tuning the parameters in Hamiltonian 6.5 at four different generations. The *red points* are samples examined by the optimizer in that generation

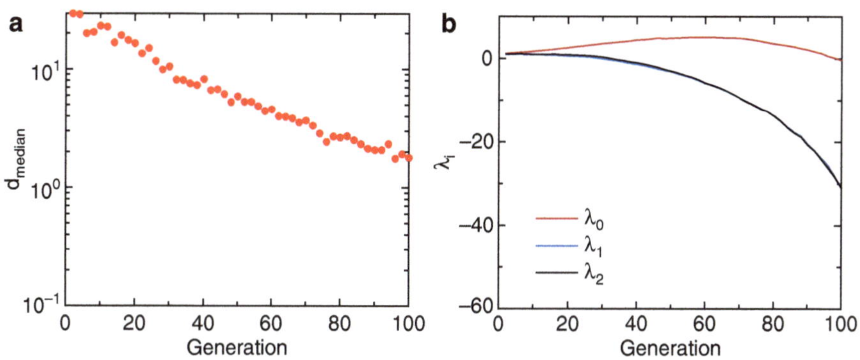

Fig. 6.4 (**a**) Median distance to the target point at (5,5) vs. generation number for a optimizer tuning an energy landscape composed of Bessel functions. Much like the CMA-ES algorithm in Chap. 2, the algorithm achieves exponential convergence after quickly exploring the search landscape. The coupling constants over time are plotted in (**b**). Note that λ_0 varies non-monotonically: the algorithm first increases the contribution from the zeroth order Bessel function and then sends it to a large negative value

6.4 Folding a Toy Protein

6.4.1 Problem Definition

The formula for self-assembly is ostensibly simple. Take a collection of thermalized, weakly interacting elements and design the interactions so that when the temperature is lowered, they spontaneously order into a desired state. But in practice, the phrase "spontaneously order" is a catch. Depending on how and which interactions are present, in all but very simple systems, different undesired states can result. This can be an Achilles heel for the self-assembly paradigm: while it might be easy to specify types of interactions theoretically or tune them experimentally, its hard to know how they should be designed put a system into a complex, desired state.

As a toy problem consider a simplified model for a protein: a string of hard, colored balls interconnected by linear springs. The balls are weakly attractive, and the interactions strengths between each are determined by the colors. For example, red and blue may be attracted more strongly than blue and green.

Fig. 6.5 The shape that results when setting every interaction strength to 100 kT. The presence of springs deforms the ground state away from an octahedral configuration

In principle, by tuning the color interactions correctly, we may be able to produce specific, nontrivial shapes. As a concrete example, suppose the chain has 6 particles and we want to form an octahedron. How should the interactions be set to make this the ground state? One solution would be to set all the interaction energies to large values compares to the spring stiffness. However, the goal here is find a set of interaction values small, or comparable to spring energies that still yields an octahedron. This result is clearly more nuanced. In fact simply setting every $\lambda = 100$ yields the shapes shown in Fig. 6.5, which clearly does not meet the design goal.

6.4.2 Simulation Methodology

We split the full Hamiltonian for this system into three parts, and the first two parts cannot be controlled by the optimizer. The first of these is an infinite hard sphere potential which prevents particles from overlapping. Second we have the spring contribution from the chain, which has a rest length that fixes particles to just touch:

$$H_s = \sum_{i=0}^{5} -\frac{1}{2}(d - r_{i,i+1})^2/(d\sigma)^2 \tag{6.6}$$

In this test problem, we fixed $d = 1.0$ and $\sigma = 0.1$, so as to make the spring energy large. The third part of the Hamiltonian is the short-ranged interaction energy, modeled as a Lennard-Jones potential

$$H_i = \sum_{ij} 4\lambda_{ij}((\frac{d}{r_{ij}})^{12} - (\frac{d}{r_{ij}})^6) \tag{6.7}$$

The λ_{ij} set the depth of the well are the control parameters for our evolutionary algorithm. In principle, they can be set to either positive or negative values

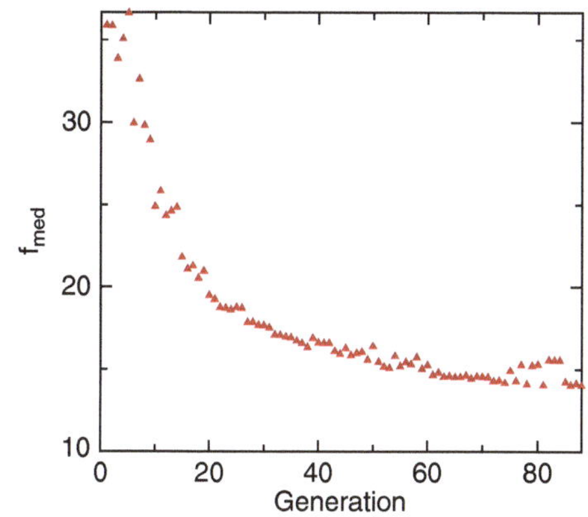

Fig. 6.6 The sum of the squared distances to the center of mass plotted against generation for the protein folding problem. In principle, an ideal octahedron made from unit radius hard spheres would have a minima of 12.0. However, the optimizer will not be able to attain this goal since the attractive Lennard-Jones potential minima does not occur when the particles make contact. Thus a problem solution should fall between 12 and $12(2)^{1/3} \approx 15$. The optimized result falls nicely between these bounds

We took the chain to be in equilibrium, so to update λ_{ij} using Eq. 2.19, we drew samples from the canonical ensemble. We again used a Metropolis-Hastings algorithm to produce samples. At each iteration, we generated 100 samples by thinning down 100,000 proposal-acceptance iterations. Proposals were drawn by adding Gaussian noise to individual particle locations. The noise standard deviation was set to $dmax[\frac{1}{\sqrt{\lambda_i}}, \sigma]$. Additionally, we discarded the first 10,000 samples, under the assumption that they remain correlated with the previous parameter settings.

Since an octahedron minimizes the sum of the distances to the center of mass, we used this as the raw fitness function. The coupling constants are again updated according to an Euler approximation to Eq. 2.19, and the process was iterated.

6.4.3 Folding an Octahedron

Figure 6.6 shows the median fitness value for each generation while Fig. 6.7 shows a typical chain configuration from each generation. Early on, the coupling constants are initialized to 0.1 kT, and so we find random, chain like configurations. However, as interaction energies grow stronger, the shapes become more compact, and more structured. Concordantly, the fitness again decreases, roughly exponentially. Ultimately after about 100 generations, the optimizer arrives at a good approximation to an octahedron.

In plotting the ultimate values of the interaction strengths, we find the optimizer arrives at an intriguingly complicated solution. Specifically, it splits the 15 total

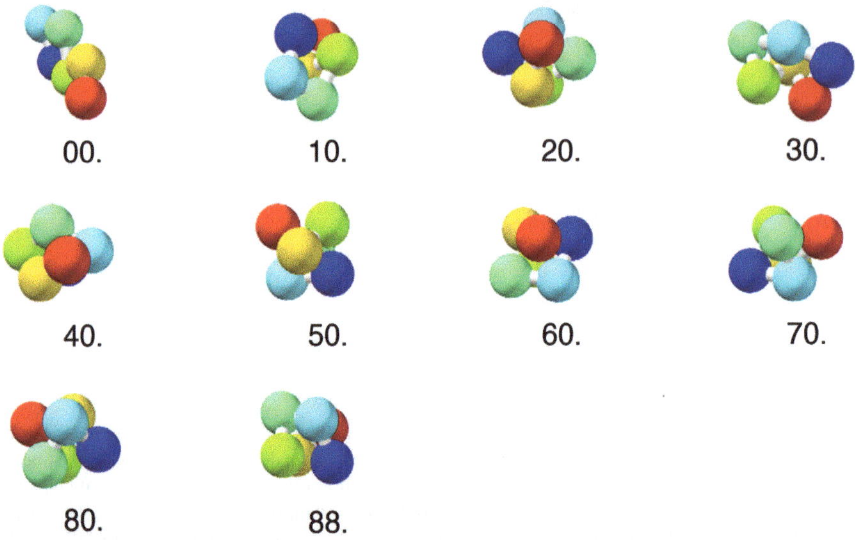

Fig. 6.7 Typical configurations of the particle chains every 10 generations. Notice that even in early generations, the optimizer is able to find configurations reminiscent of the octahedron, although the coupling constants are still relatively weak compared to the spring strengths

interaction parameters into groups (Fig. 6.9). The largest coupling constant group bonds the first (red) and last (dark blue), second (yellow) and fourth (cyan), and third (green) and fifth (light blue) and third particles in the chain. Further, by plotting the coupling constants over time, we find the differentiation between these groups occurs at roughly 70 generations (Fig. 6.8). Further we note that all of the coupling constants between neighboring elements in the chain are given the same negative value, acting to stretch the chain out. The results are plotted as a network in Fig. 6.10. We anticipated that the bond interaction strengths would have resembled the contact network for an octahedron. Instead the optimizer selects a different, but nevertheless structured parameter set, that still meets the design goal.

6.5 Summary

In any given simulation, there is an extraordinary amount of information contained in the statistics. So much so, that in defining optimization problems we typically discard most of it to make the problem comprehensible. Yet by doing so, we believe that the employed optimizer is neither making full use of computer effort used to generate simulation results, nor the physical knowledge behind them.

The results here show how to start addressing this issue with artificial evolution as a formalism. The resulting method is surprisingly easy to implement and adjust. The only ingredients are a simulation engine that produces statistical data, a

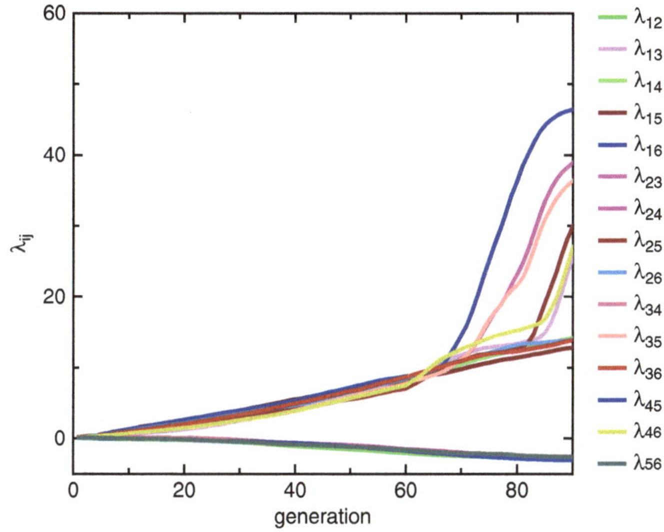

Fig. 6.8 Time evolution of all 15 coupling constants in the polymer chain. Note that nearest neighbor interactions are all set to small negative values, and pairs of particular interactions seem to adjust in coordination, even though this behavior was not pre-specified in constructing the optimizer

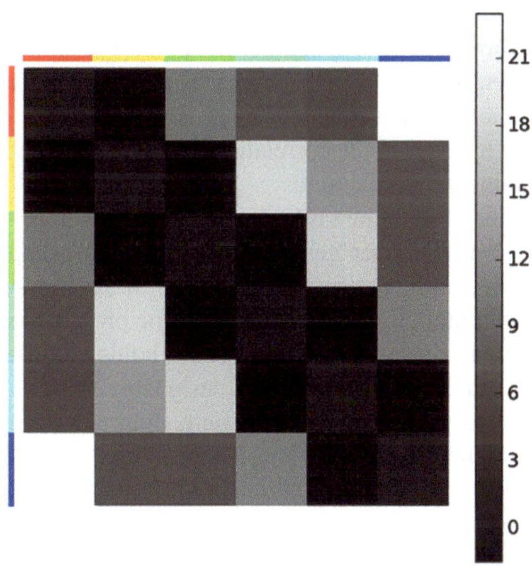

Fig. 6.9 Energy interaction matrix the optimizer constructed to fold the six particle chain into an octahedron. The x and y axis on the matrix are ordered with respect to the particle position on chain and the colors correspond to the ball colors plotted in Fig. 6.7. Thus the (1,2) position in the matrix is the interaction energy between the first and second particles. The matrix is surprisingly structured: no particles interact strongly with their immediate neighbors and the interactions that are ostensibly grouped

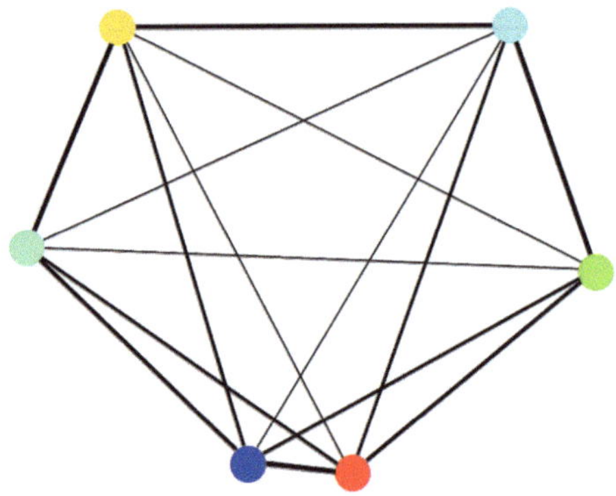

Fig. 6.10 The interaction matrix plotted as a network, where the thickness of each bond is proportional to the logarithm of the interaction strength. Colors correspond to the ball colors in Fig. 6.7. We find that the geometry of the chain, rather than the octahedron, is prevalent in several of the features. For instance, no two neighbors in the chain feel strong interactions and the tightest bonding pair is between the ends of the chain

distributional form that encapsulates the data, and a design goal. The optimizer can then be left to run, interacting dynamically with the simulations to produce good solutions.

Two features make this approach uniquely attractive. First, the algorithm iterates using black-box information. For example, in the case explored here the energy terms could have been the results of complicated computer calculations, or just as easily, direct experimental measurements. In principle, the code could even be run in the lab by directly measuring the correlation functions and leaving the algorithm to tune physical parameters in real time.

Second the optimization program is integrated tightly with the simulation engine, requiring just one simulation per generation. By contrast, running the CMA-ES directly to design the coupling constants in any of these problems would require 10 time as much computation effort per update. For problems where simulations are computational intensive, an integrated approach holds promise as a means to the speedups in runtime that make new materials designable.

Looking beyond the preliminary results here, future work may find ways to extend this framework beyond equilibrium. To the extent that a materials system is statistical and that the underlying distribution can be parameterized, an evolutionary algorithm can always be constructed that updates the parameters to solve a design problem. Thus non-equilibrium processes, provided they have a statistical mechanics description, are fair game. For example, if we simulate a diffusion process by adding Gaussian noise to a mean drift, then the paths are distributed by a product of Gaussian distributions conditioned on the prior steps. Thus, the paths are statistical objects, with diffusion and drift as the distribution parameters. One could then build an optimizer that tunes these time varying functions, thereby evolving processing protocols and interactions for an out of equilibrium system with the same formalism.

References

Dahiyat, B. I., & Mayo, S. L. (1997). De novo protein design: Fully automated sequence selection. *Science, 278*(5335), 82–87.

Jain, A., Bollinger, J. A., & Truskett, T. M. (2014). Inverse methods for material design. *American Institute of Chemical Engineers Journal, 60*(8), 2732–2740.

Kuhlman, B., Dantas, G., Ireton, G. C., Varani, G., Stoddard, B. L., & Baker, D. (2003). Design of a novel globular protein fold with atomic-level accuracy. *Science, 302*(5649), 1364–1368.

Oganov, A. R., Lyakhov, A. O., & Valle, M. (2011). How evolutionary crystal structure prediction works-and why. *Accounts of Chemical Research, 44*(3), 227–237.

Ollivier, Y., Arnold, L., Auger, A., & Hansen, N. (2011). Information-geometric optimization algorithms: A unifying picture via invariance principles. arXiv preprint arXiv:1106.3708.

Torquato, S. (2009). Inverse optimization techniques for targeted self-assembly. *Soft Matter, 5*(6), 1157–1173.

Chapter 7
Conclusions

The central tenet of condensed matter physics is that all the emergent behavior that defines a material at the macro-scale is linked back to some kind of microscopic roots. The level of magnification needed to describe a property may vary from atoms to grains of sand, but there is always a causal relationship that points from the micro-scale to the macro-scale. Expressing this dichotomy poses two intrinsic possibilities for study: find relationships that link causes to effects and find relationships that take effects into causes. The latter is what we call design.

As the prior chapters show, design problems, at least for materials systems, both come in a wide variety and can generally be solved by machines. In the simplest case, there is optimization, in which the design task is to find one cause that maximizes a given effect. Slightly more general than that, we considered inverting relationships, where we sought out a rule that transforms an arbitrary effect into the appropriate cause. We've examined the process of designing as a function that takes simulation settings and goals as inputs, as well as an operation that a computer can actively perform on a material system. To us, it is encouraging that these issues could be successfully examined for complicated far from equilibrium systems with an automated, and largely generic, framework.

These proof of principle problems hint at the risks and rewards associated with perusing the idea of automated materials design. The most striking benefit is that previously intractable problems in materials science can start to be addressed. By doing so, the complexities that prohibit traditional engineering tasks can be leveraged to make novel or nuanced materials. Case in point are the optimization results for granular mechanics in Chap. 3. To our knowledge, there is no prior work in optimizing the mechanical response of granular materials through the use of shape. Yet, we showed how an integrated computational-experimental approach can be used to fine tune variations in response, like the changes between the stiffest and softest packing shapes. In principle, this method could make at least granular materials conform as closely as possible to any physically realizable demand for aggregate stiffness. We also showed how the same algorithm can design a strain

M.Z. Miskin, *The Automated Design of Materials Far From Equilibrium*,
Springer Theses, DOI 10.1007/978-3-319-24621-5_7

stiffening particle. To produce this qualitative change in how the packing resists strain, the algorithm must have found a way to exploit the nonlinearities that make explaining granular mechanics so difficult. Taken together, these results demonstrate precise control and gross variation using one kind of machinery.

Automated design systems also invite new ways of thinking about materials since they allow one to take the process of design as granted. For example, in Chaps. 4 and 5, we took the fact that we could seek out desired microscopic features as given. This allows the materials designer to be imaginative or inquisitive about new aspects of the design problem. One can work to invent design rules that generically solve hard inverse problems. Or, one can classify types of designs and hunt down new physics at the boundaries between them. These are some of the exciting new perspectives that begin to appear when thinking about materials from the top down: the user can specify aspects of a material is abstractly in terms of functionalities, offloading the task of realizing those functionalities on the computer.

One could criticize that there are risks to automated design arising from the uncertainties associated with it. For example, it is usually impossible to tell if the objects returned by the optimizer are in fact ideal. Alternatively, the results of any optimization are contained within the physics of the foreword simulation engine. Thus, automated design cannot, by itself, discover new material properties if they are the consequence of completely new physics.

We have to accept the fact that the result of any optimization is not known to be the best: it is simply a hypothesis of what object might be best. But arguably this is always the state of affairs in materials science. The natural way to address this concern is to use design results as the starting point for scientific inquiry, and weigh the evidence for or against them. For example, in Chaps. 3 and 4, we did this by considering additional, but similar design problems (e.g. densest/loosest), and by experimenting on real world prototypes. In Chap. 4, we even took the next step to test a new hypothesis that rationalizes the results. In other words, starting with the optimization results as a hypothesis we moved through the first few steps of the scientific method.

From this perspective, automated materials design fits neatly in the grand scheme of materials science. Artificial evolution, in particular the type described in Chap. 2, is a tool for hypothesis generation through the empirical synthesis of simulation data. We even played on this idea in Chap. 6 where we designed an algorithm to act like an experimentalist performing exploratory research by turning the knobs that control a virtual experiment. This kind of radical empiricism is what makes artificial evolution a particular good method for creating concepts ab initio. The machine has no real world prejudices, biases, or innate misconceptions. Instead its "motivation" for picking one design over another is driven purely by simulation data. This makes it prime for exploring systems where little, if any, physics is known. Ultimately, we see science and design as complimentary: while design moves from macro to micro, the scientific method goes back the other way, allowing the results of one to strengthen and inform the other.

For materials far from equilibrium, this cycle of feedback is particularly auspicious. Far from equilibrium is uncharted territory for physics, making its study as much an intellectual affair as a practical one. Without having to make substantial changes, we can envision extending the work here to operate on colloids, foams and molecular glasses. Our work invites answering questions like "what processing condition optimizes stiffness in a metallic foam?" or "what shapes in a suspension shear thicken strongest?" Like our studies in granular materials, optimizing and tailoring these materials, should spark sharp questions about what gives each special properties, lending to deep physical insight. Thus, we encourage future work to focus on the hard systems far from the equilibrium state because it is here that the true power of automated design is the most apparent: it is far from equilibrium where the seeds of knowledge and the seeds of technology get to be discovered at the same time.

MIX
Papier aus verantwortungsvollen Quellen
Paper from responsible sources
FSC® C105338

If you have any concerns about our products,
you can contact us on
ProductSafety@springernature.com

In case Publisher is established outside the EU,
the EU authorized representative is:
Springer Nature Customer Service Center GmbH
Europaplatz 3, 69115 Heidelberg, Germany

Printed by Libri Plureos GmbH
in Hamburg, Germany